SpringerBriefs in Physics

SpringerBriefs in Physics are a series of slim high-quality publications encompassing the entire spectrum of physics. Manuscripts for SpringerBriefs in Physics will be evaluated by Springer and by members of the Editorial Board. Proposals and other communication should be sent to your Publishing Editors at Springer.

Featuring compact volumes of 50 to 125 pages (approximately 20,000–45,000 words), Briefs are shorter than a conventional book but longer than a journal article. Thus, Briefs serve as timely, concise tools for students, researchers, and professionals. Typical texts for publication might include:

- A snapshot review of the current state of a hot or emerging field
- A concise introduction to core concepts that students must understand in order to make independent contributions
- An extended research report giving more details and discussion than is possible in a conventional journal article
- A manual describing underlying principles and best practices for an experimental technique
- An essay exploring new ideas within physics, related philosophical issues, or broader topics such as science and society

Briefs allow authors to present their ideas and readers to absorb them with minimal time investment. Briefs will be published as part of Springer's eBook collection, with millions of users worldwide. In addition, they will be available, just like other books, for individual print and electronic purchase. Briefs are characterized by fast, global electronic dissemination, straightforward publishing agreements, easy-to-use manuscript preparation and formatting guidelines, and expedited production schedules. We aim for publication 8–12 weeks after acceptance.

More information about this series at http://www.springer.com/series/8902

Marco Fabbrichesi · Emidio Gabrielli ·
Gaia Lanfranchi

The Physics of the Dark Photon

A Primer

 Springer

Marco Fabbrichesi
Sezione di Trieste
INFN
Trieste, Italy

Emidio Gabrielli
Department of Physics
University of Trieste
Trieste, Italy

Gaia Lanfranchi
National Laboratory of Frascati
Frascati, Italy

ISSN 2191-5423   ISSN 2191-5431   (electronic)
SpringerBriefs in Physics
ISBN 978-3-030-62518-4   ISBN 978-3-030-62519-1   (eBook)
https://doi.org/10.1007/978-3-030-62519-1

This Springer imprint is published by the registered company Springer Nature Switzerland AG
The registered company address is: Gewerbestrasse 11, 6330 Cham, Switzerland

Preface

New Particles Beyond the Standard Model have always been thought to be charged under at least some of the same gauge interactions of ordinary particles. Although this assumption has driven the theoretical speculations, as well as the experimental searches of the last 50 years, it has also been increasingly challenged by the negative results of all these searches—and the mounting frustration for the failure to discover any of these hypothetical new particles.

As the hope of a breakthrough along these lines is waning, interest in a *dark sector*—dark because not charged under the Standard Model gauge groups—is growing: Maybe no new particles have been seen simply because they do not interact through the Standard Model gauge interactions.

The dark sector can be simple or have a complex structure with many states—some of which can be dark matter candidates.

If the dark and the visible sectors were to interact only gravitationally—which they cannot avoid—there would be little hope of observing in the laboratory particles belonging to the dark sector. A similar problem exists for dark matter: Although its presence is motivated by gravitational physics, it is searched mostly through its putative weak interactions—as in the direct- and indirect-detection searches of a weakly interacting massive particle. For the same reason, we must pin our hopes on assuming that dark and ordinary sectors also interact through a *portal*—as the current terminology has it—that is, through a sallow glimmer, in a manner that, though feeble, is (at least in principle) experimentally accessible.

The portal may take various forms that can be classified by the type and dimension of its operators. The best motivated and most studied cases contain relevant operators taking different forms depending on the spin of the mediator: *Vector* (spin 1), *Neutrino* (spin 1/2), *Higgs* (scalar) and *Axion* (pseudo-scalar).

Among these possible portals, the vector portal is the one where the interaction takes place because of the kinetic mixing between one dark and one visible Abelian gauge boson (nonAbelian gauge bosons do not mix). The visible photon is taken to be the boson of the $U(1)$ gauge group of electromagnetism—or, above the electroweak symmetry-breaking scale, of the hyper-charge—while the *dark photon* comes to be identified as the boson of an extra $U(1)$ symmetry.

The concept of a portal—which at first blush might seem rather harmless—actually represents a radical departure from what is the main conceptual outcome of our study of particle physics, namely, the gauge principle and idea that all interactions must be described by a gauge theory. The portal, and the new interactions that it brings into the picture, adds a significant exception to this principle. Among the possible portals, the vector case deviates the least from the gauge principle as it only introduces a mixing for the gauge bosons while the interaction to matter remains of the gauge type (albeit with an un-quantized charge). Instead, the other kinds of portal imply a manifest new violation to the gauge principle, the other's being the notable case of the Yukawa and self-interactions of the Higgs boson—which are themselves, exactly because of their not being gauge interactions, the least understood part of the Standard Model.

There is an additional and important reason to study the dark sector in general, and the dark photon in particular: The main motivation in introducing new-physics scenarios is to use them as a foil for the Standard Model in mapping possible experimental discrepancies. In the absence of clearly identified new states, the many parameters, for instance, of the supersymmetric extensions to the Standard Model or even of the effective field theory approach to physics beyond the Standard Model, are working against their usefulness. Instead, each dark sector can be reduced to few parameters—to wit, just two in the case of the dark photon—in terms of which the possible discrepancies with respect to the Standard Model are more effectively mapped in the experimental searches and the potential discovery more discernible.

Trieste, Italy   Marco Fabbrichesi
Genève, Switzerland   Emidio Gabrielli
October 2020   Gaia Lanfranchi

Acknowledgements

The digital inclusion of some of the experimental limits in the figures was done by means of WebPlotDigitizer.[1] Numerical results for some of the dark photon limits were provided by the Darkcast package.[2]

We would like to thank Tram Acuña, Andrea Celentano, Monica D'Onofrio, Angelo Esposito, Oliver Fischer, Phil Ilten, Sam McDermott, Maxim Pospelov, Diego Redigolo, Josh Ruderman, Piero Ullio, Alfredo Urbano and Mike Williams for useful discussions and suggestions.

MF is affiliated to the Physics Department of the University of Trieste and the *Scuola Internazionale Superiore di Studi Avanzati* (SISSA), Trieste, Italy. MF and EG are affiliated to the Institute for Fundamental Physics of the Universe (IFPU), Trieste, Italy. The support of all these institutions is gratefully acknowledged. EG thanks the Department of Theoretical Physics of CERN for its kind hospitality during the preparation of this work.

---

[1]Website: https://automeris.io/WebPlotDigitizer.

[2]Website: https://gitlab.com/philten/darkcast.

Contents

Chapter 1
Introduction

The fundamental interactions among the elementary particles of the Standard Model (SM) are provided by the photon, the weak gauge bosons $W^{\pm}$ and Z^0 and the gluons. These are the gauge bosons, spin 1 vector particles that make the symmetries on which the SM is based into local symmetries and the SM into a gauge theory.

The dark photon is a new gauge boson whose existence has been conjectured. The names *para-* [1], *hidden-sector, secluded photon* and *U-boson* [2] have also being used to indicate the same particle. The dark photon is dark because it arises from a symmetry of a hypothetical dark sector comprising particles completely neutral under the SM interactions. The neutrality of ordinary matter makes it blind to this new gauge boson which is accordingly invisible and therefore characterized as dark.

The idea of adding to the SM a new gauge boson similar to the photon was first considered in the context of supersymmetric theories in [3, 4]. It was discussed more in general shortly after in [5, 6].

The dark sector is assumed to exist as a world parallel to our own. It may contain few or many states, and these can be fermions or scalars or both, depending on the model. Dark matter proper—the existence of which is deemed necessary to explain astrophysical data—is found among these states. Its relic density can be computed and constrained by observational data. In addition, the dark states can interact; their interactions can be Yukawa-like or mediated by dark gauge bosons or both depending on the model.

Dark though it is, the dark photon can be detected because of its kinetic mixing with the ordinary, visible photon. This kinetic mixing is always possible because the field strengths of two Abelian gauge fields can be multiplied together to give a dimension four operator. The existence of such an operator means that the two gauge bosons can go into each other as they propagate. This kinetic mixing provides the portal linking the dark and visible sectors. It is this portal that makes possible to detect the dark photon in the experiments.

In this primer, we review the physics of this new gauge boson from the theoretical and the experimental point of view. We explain how the dark photon enters laboratory, astrophysical and cosmological observations as well as dark matter physics.

© The Author(s), under exclusive license to Springer Nature Switzerland AG 2021   1
M. Fabbrichesi et al., *The Physics of the Dark Photon*,
SpringerBriefs in Physics, https://doi.org/10.1007/978-3-030-62519-1_1

As explained in detail in Sect. 1.1, there are actually two kinds of dark photons: The massless and the massive—whose theoretical frameworks as well as experimental signatures are quite distinct. They give rise to dark sectors with different features; their characteristic physics and experimental searches are best reviewed separately. The massive dark photon has been receiving so far most of the attention because it couples directly to the SM currents and is more readily accessible in the experimental searches. The massless dark photon arises from a sound theoretical framework and, as we shall argue, provides, with respect to the massive case, a comparably rich, if perhaps more challenging, experimental target.

We look into the ultraviolet (UV) completion of models of the dark photon in Sect. 1.2 to better understand the origin of their interactions with the SM particles.

Section 1.3 describes the interplay between the dark photon and dark matter and introduces many of the definitions used in the experimental searches.

We survey the current and future experimental limits on the parameters of the massless and massive dark photons together with the related bounds on milli-charged fermions. We discuss all these constrains for the massless case in Sect. 2 and for the massive case in Sect. 3. At the best of our knowledge, these two sections provide the reader with a comprehensive review of the physics of the dark photon.

We collect in three appendices a number of definitions and equations, which the reader may find useful to better follow the discussion in the main text.

In the past few years a number of reports on the dark sector (and the massive dark photon within it) have been published [7–14]. The interested reader can therein find different points of view to complement the present review as well as additional details on the other portals. A previous discussion of the astrophysical, cosmological and other constraints for the massless dark photon can be found in [15].

1.1 Massless and Massive Dark Photons

The most general kinetic part of the Lagrangian of two Abelian gauge bosons, described by two gauge groups $U(1)_a$ and $U(1)_b$, is given by

$$\mathcal{L}_0 = -\frac{1}{4} F_{a\mu\nu} F_a^{\mu\nu} - \frac{1}{4} F_{b\mu\nu} F_b^{\mu\nu} - \frac{\varepsilon}{2} F_{a\mu\nu} F_b^{\mu\nu} . \tag{1.1}$$

The gauge boson A_b^μ is taken to couple to the current J_μ of ordinary SM matter, the other, A_a^μ, to the current J'_μ, which is made of dark-sector matter, to give the Lagrangian

$$\mathcal{L} = e J_\mu A_b^\mu + e' J'_\mu A_a^\mu , \tag{1.2}$$

with e and e' the respective coupling constants.

To discuss the physics arising from the Lagrangians in Eqs. (1.1) and (1.2), it is useful to identify from the very beginning two kinds of dark photons:

– the massless kind, which, as we are about to show, does not couple directly to any of the SM currents and interacts instead with ordinary matter only through operators of dimension higher than four;
– the massive kind, which couples to ordinary matter through a current (with arbitrary charge), that is, a renormalizable operator of dimension four. The massless limit of this case does not correspond to the massless case above.

Because of their different coupling to SM particles, the two kinds are best discussed separately.

Let us first consider the massless case.

As first discussed in [1] in this case the classical Lagrangian can be diagonalized. What happens at the quantum level and how the mixing manifests itself has been analyzed in detail in [16] for the unbroken gauge theory as well as the spontaneously broken case (see, also, the appendix of [17] which we mostly follow).

The kinetic terms in Eq. (1.1) can be diagonalized by rotating the gauge fields as

$$
\begin{pmatrix} A_a^\mu \\ A_b^\mu \end{pmatrix} = \begin{pmatrix} \dfrac{1}{\sqrt{1-\varepsilon^2}} & 0 \\ -\dfrac{\varepsilon}{\sqrt{1-\varepsilon^2}} & 1 \end{pmatrix} \begin{pmatrix} \cos\theta & -\sin\theta \\ \sin\theta & \cos\theta \end{pmatrix} \begin{pmatrix} A'^\mu \\ A^\mu \end{pmatrix},
\tag{1.3}
$$

where now we can identify A^μ with the ordinary photon and A'^μ with the dark photon. The additional orthogonal rotation in Eq. (1.3) is always possible and introduces an angle θ which is arbitrary as long as the gauge bosons are massless.

After the rotation in Eq. (1.3), the interaction Lagrangian in Eq. (1.2) becomes

$$
\mathcal{L}' = \left[\frac{e'\cos\theta}{\sqrt{1-\varepsilon^2}} J'_\mu + e\left(\sin\theta - \frac{\varepsilon\cos\theta}{\sqrt{1-\varepsilon^2}} \right) J_\mu \right] A'^\mu
$$
$$
+ \left[-\frac{e'\sin\theta}{\sqrt{1-\varepsilon^2}} J'_\mu + e\left(\cos\theta + \frac{\varepsilon\sin\theta}{\sqrt{1-\varepsilon^2}} \right) J_\mu \right] A^\mu.
\tag{1.4}
$$

By choosing $\sin\theta = 0$ $(\cos\theta = 1)$ (see right-side of Fig. 1.1) we can have the ordinary photon A_μ coupled only to the ordinary current J_μ while the dark photon couples to both the ordinary and the dark current J'_μ, the former with strength $\varepsilon e/\sqrt{1-\varepsilon^2}$ proportional to the mixing parameter ε. The Lagrangian is therefore:

$$
\boxed{\mathcal{L}' = \left[\frac{e'}{\sqrt{1-\varepsilon^2}} J'_\mu - \frac{e\varepsilon}{\sqrt{1-\varepsilon^2}} J_\mu \right] A'^\mu + e J_\mu A^\mu.}
\tag{1.5}
$$

Vice versa, with the choice $\sin\theta = \varepsilon$ and $\cos\theta = \sqrt{1-\varepsilon^2}$ (see left-side of Fig. 1.1), we have the opposite situation with the dark photon only coupled to the dark current and the ordinary photon to both currents, with strength $\varepsilon e/\sqrt{1-\varepsilon^2}$ to the dark one. This latter coupling between the dark-sector matter to the ordinary photon is called a *milli-charge*. Its value is experimentally known to be small [18].

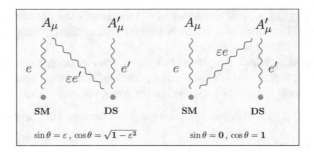

Fig. 1.1 Scheme of the coupling of the ordinary (A_μ) and dark (A'_μ) photon to the SM and dark-sector (DS) particles for the two choices of the angle θ discussed in the main text. e and e' are the couplings of the ordinary and dark photons to their respective sectors

The dark photon sees ordinary matter only through the effect of operators like the magnetic moment or the charge form factors (of dimension higher than four). This is the choice defining the massless dark photon proper:

$$\mathcal{L}' = e' J'_\mu A'^\mu + \left[-\frac{e'\varepsilon}{\sqrt{1-\varepsilon^2}} J'_\mu + \frac{e}{\sqrt{1-\varepsilon^2}} J_\mu \right] A^\mu \tag{1.6}$$

If the gauge symmetry is spontaneously broken, the diagonalization of the mass terms locks the angle θ to the value required by the rotation of the gauge fields to the mass eigenstates and we cannot have that one of the two currents only couples to one of the two gauge bosons.

This is also the case when the $U(1)$ gauge bosons acquire a mass by means of the Stueckelberg Lagrangian (see [19] for a review and the relevant references)

$$\mathcal{L}_{Stu} = -\frac{1}{2} M_a^2 A_{a\mu} A_a^\mu - \frac{1}{2} M_b^2 A_{b\mu} A_b^\mu - M_a M_b A_{a\mu} A_b^\mu. \tag{1.7}$$

In this case, as in the spontaneously broken case, the angle θ is fixed and equal to

$$\sin\theta = \frac{\delta\sqrt{1-\varepsilon^2}}{\sqrt{1-2\delta\varepsilon+\delta^2}} \qquad \cos\theta = \frac{1-\delta\varepsilon}{\sqrt{1-2\delta\varepsilon+\delta^2}} \tag{1.8}$$

where $\delta = M_b/M_a$, and we have no longer the freedom of rotating the fields as in Eq. (1.3). The Lagrangian in Eq. (1.4) is now

$$\mathcal{L}'' = \frac{1}{\sqrt{1-2\delta\varepsilon+\delta^2}} \left[\frac{e'(1-\delta\varepsilon)}{\sqrt{1-\varepsilon^2}} J'_\mu + \frac{e(\delta-\varepsilon)}{\sqrt{1-\varepsilon^2}} J_\mu \right] A'^\mu$$

$$+ \frac{1}{\sqrt{1-2\delta\varepsilon+\delta^2}} \left[e J_\mu - \delta e' J'_\mu \right] A^\mu. \tag{1.9}$$

The case of spontaneously broken symmetry can be distinguished from the Stueckelberg mass terms because the former will give rise to processes in which the dark photon is produced together with the dark Higgs boson, the vacuum expectation value of which hides the symmetry.

Whereas the Lagrangian in Eq. (1.9) is the most general, the simplest and most frequently discussed case consists in giving mass directly to only one of the $U(1)$ gauge bosons so that, for instance, $M_b = 0$ in Eq. (1.7), the mass states are already diagonal. Even in this simple case, the mass term removes the freedom of choosing the angle θ in Eq. (1.3). With this choice, $\delta = 0$ in Eq. (1.9), the ordinary photon couples only to ordinary matter and the massive dark photon is characterized by a direct coupling to the electromagnetic current of the the SM particles (in addition to that to dark-sector matter) and described by the Lagrangian

$$\mathcal{L} \supset -\frac{e\varepsilon}{\sqrt{1-\varepsilon^2}} J_\mu A'^\mu \simeq -e\,\varepsilon\, J_\mu A'^\mu\,, \tag{1.10}$$

as in Eq. (1.5) above. This is the choice defining the massive dark photon. The coupling of the massive dark photon to SM particles is not quantized—taking the arbitrary value $e\varepsilon$. Because of this direct current-like coupling to ordinary matter, it is the spontaneously broken or Stueckelberg massive dark photon that is mostly discussed in the literature and considered in the experimental proposals.

Notice that the massive dark photon has the same couplings as the massless dark photon after choosing $\sin\theta = 0$ (right-side of Fig. 1.1); this case therefore represents the limit of vanishing mass of the massive dark photon. On the contrary, the massless dark photon proper—corresponding to the choice $\tan\theta = \left[\varepsilon/\sqrt{1-\varepsilon^2}\right]$— is not related to any limiting case of the massive dark photon.

There are no electromagnetic milli-charged particles in the massive case; they are present only if both $U(1)$ gauge groups are spontaneously broken (or equivalently $M_b \neq 0$ in the Stueckelberg Lagrangian in Eq. (1.7))—which is not the case of our world where the photon is massless.

1.1.1 Kinetic Mixing: Electric or Hyper-Charge?

There seems to be the choice in the kinetic mixing in Eq. (1.1) between the $U(1)_{e.m.}$ group of electric charge and the $U(1)_Y$ group of the hyper-charge, with mixing parameter ε defined as in Eq. (1.1). Concerning the massless dark photon, these two choices give rise to the same physics, since the dark photon remains decoupled from the SM fields at the tree-level. The only difference is that the photon and Z-boson are now both coupled to the dark-sector current, with $e_D\varepsilon/\sqrt{1-\varepsilon^2}\cos\theta_W$ and $e_D\varepsilon/\sqrt{1-\varepsilon^2}\sin\theta_W$ strength, respectively.

Let now consider the massive dark-photon coupling to hyper-charge. In this case it is convenient to parametrize the coupling of the dark photon to the hyper-charge as

$$\tilde{\mathcal{L}} = -\frac{\varepsilon}{2 \cos \theta_W} \tilde{F}'_{\mu\nu} B^{\mu\nu} . \qquad (1.11)$$

The usual diagonalization of the gauge bosons W^3_μ and B_μ now includes also the dark photon $\tilde{A}'_\mu$ (in the non-diagonal basis) so that the physical gauge bosons Z_μ and A_μ also contain a dark-photon component A'_μ in the mass eigenstate basis. In particular, at the $O(\varepsilon)$ in the expansion, we have

$$\begin{pmatrix} W^3_\mu \\ B_\mu \\ \tilde{A}'_\mu \end{pmatrix} = \begin{pmatrix} c_W & s_W & -s_W \varepsilon \\ -s_W & c_W & -c_W \varepsilon \\ t_W \varepsilon & 0 & 1 \end{pmatrix} \begin{pmatrix} Z_\mu \\ A_\mu \\ A'_\mu \end{pmatrix} , \qquad (1.12)$$

where c_W, s_W and t_W are the usual cosine, sine, and tangent of the Weinberg angle θ_W, respectively. New couplings of the massive dark photon to the SM fermions appear for the photon and the Z gauge boson up to $O(\varepsilon^2)$:

$$\mathcal{L} \supset -e \varepsilon J^\mu A'_\mu + e' \varepsilon t_W J'^\mu Z_\mu + e' J^{\mu\prime} A'_\mu , \qquad (1.13)$$

where J_μ is the EM current, while J'_μ and e' are the matter current and coupling of the dark-photon in the dark sector, respectively. After integrating out the Z boson, we see that the coupling of the massive dark photon to the SM fermions is recovered as $-e\varepsilon$.

Which coupling is used depends then only on the energy of the processes considered, with the direct coupling to the photon for all processes below the electroweak scale breaking, and the hyper-charge above it. Since all limits are to be considered approximately within the order of magnitude, the presence of the factor c_W in the definition in Eq. (1.11) does not matter. The Lagrangian in Eq. (1.13) shows that, if the mixing is between the dark photon and the hyper-charge, the Z gauge boson acquires a milli-charged coupling strength $e' t_W \varepsilon$ to the dark sector current.

For completeness, let us also recall two other possibilities that have been discussed in the literature:

- There is no kinetic mixing as in Eq. (1.1) but the mass term between the dark photon and the Z-boson is taken non-diagonal and therefore giving a mixing between these two states [20–24]. The dark photon is named the dark Z and there are characteristic experimental signatures in parity violating processes and the coupling to neutrinos;
- The $B - L$ global symmetry (or other conserved flavor symmetries) are gauged and taken to be the $U(1)$ group of the dark photon, which mixes with the hyper-charge [25–27]. There is direct coupling to the SM fermions in this case and the dark photon is no longer dark.

Although their implementation is not discussed in this review, other interesting generalizations—as, for instance, the dark photon to be considered a Kaluza-Klein state in a model with large extra-dimensions [28] or the interplay between the neutrino see-saw mechanism and the dark photon [29]—should be borne in mind.

1.1.2 Embedding in a NonAbelian Group

In the massless case, the ordinary photon still couples to the dark sector with a milli-charge εe. As reviewed in the next section, there are very stringent limits on the size of such a milli-charge, at least for reasonably light dark states. To avoid the necessity of assuming a very small milli-charge, one can assume that the dark $U(1)$ group is a symmetry left over after the spontaneous breaking of a larger nonAbelian group.

The simplest realization of this symmetry breaking is provided by the group $SU(2)$ spontaneously broken to $U(1)$ by the vacuum expectation value of the neutral component of a scalar field in the adjoint representation.

In this scenario, the mixing term in Eq. (1.1) cannot be written because the larger group has traceless generators. The absence of mixing is in this case protected against radiative corrections and the dark and the ordinary photons see only their respective sectors (at least through renormalizable operators).

This scenario is also suggested by the extra Landau pole that otherwise would be present—assuming that the Landau pole of the ordinary $U(1)$ is removed by the embedding of the SM in a scenario of grand unified theory.

If we assume that the dark photon arises from a nonAbelian group, there is no milli-charged coupling of the dark sector to ordinary photons. On the other hand, all states in the dark sector must come as multiplets of the nonAbelian group and the possible experimental signatures of this additional structure can be searched for.

1.2 UV Models

Because the massive dark photon couples directly to the SM electromagnetic current, its phenomenology is rather independent of the details of the underlaying UV completion. The two parameters ε and $m_{A'}$ suffice to fully describe the experimental searches.

The case of the massless dark photon is more complicated because the coupling to the SM particles only takes place through higher order operators whose structure heavily depends on the underlaying UV model. Even though it is possible to frame the experimental search in terms of the effective scale of these operators (as we do in Sect. 2), the limits thus found begs to be framed in terms of the UV model parameters, namely the masses and the coupling of the dark sector states, in addition to the dark photon itself. For this reason, it is useful in this case to introduce a minimal UV model (as we do in Sect. 2.3) to provide the relationships among the parameters of the model and thus possible to relate different limits that are instead independent or not present under the portal interaction.

1.2.1   Massive Dark Photon: Origin and Size of the Mixing Parameter

The size of the mixing parameter ε is arbitrary. It is this feature that makes the charge not quantized. At the same time, it cannot be $O(1)$ because, if so, the massive dark photon would have already been discovered.

A natural suppression of ε is achieved if the mixing only comes as a correction at one- or two-loop level in some UV completion. This is achieved in a natural manner if the tree-level mixing is set to zero. One looks for the renormalization of the model and introduces the necessary counter-terms, of which the mixing in Eq. (1.1) is one. If there are states in the UV completion carrying both ordinary and dark charges, the loop of these states generates the mixing but it comes suppressed by the loop factor (neglecting logarithmic terms) and therefore of order, say, $1/(16\pi^2)$ times the square of the coupling constant and therefore approximately $O(10^{-3})$, for a perturbative value of such a coupling. One can further suppress such a term by assuming that the states carrying both charges come in doublets of opposite dark charges. In this case, the first contribution is at the two-loop level, and approximately of order $O(10^{-5})$. If the mixing originates in the exchange of heavy messenger fields [30] or in a multi-loop contribution [31, 32], its value can be smaller.

Even smaller values of the parameter ε are expected if the origin of the mixing is non-perturbative; for example, values between $O(10^{-12})$ and $O(10^{-6})$ have been discussed—mostly within the broad heading of string compactification [33–38], or in scenarios of SUSY breaking [39] and hidden valley [40]. These arguments are often cited to motivate experimental searches in the region of small mixing parameter ε in the case of the massive dark photon—regardless of the large uncertainties in the predictions of the corresponding theoretical approaches.

1.2.2   Massless Dark Photon: Higher-Order Operators

The massless dark photon does not interact directly with the currents of the SM fermions. The higher-order operators through which the interaction with ordinary matter ψ^i takes place start with the dimension-five operators in the Lagrangian

$$\mathcal{L} = \frac{e_D}{2\Lambda_5} \overline{\psi}^i \, \sigma_{\mu\nu} \left(\mathbb{D}_M^{ij} + i\gamma_5 \, \mathbb{D}_E^{ij} \right) \psi^j \, F'^{\mu\nu} \,, \tag{1.14}$$

where $F'_{\mu\nu}$ is the field strength associated to the dark photon field A'_μ, and $\sigma_{\mu\nu} = i/2\,[\gamma_\mu, \gamma_\nu]$. The operator proportional to the coefficient $\mathbb{D}_M$ is the magnetic dipole moment and that proportional to the coefficient $\mathbb{D}_E$ is the electric dipole moment. The indices i and j in the fermion fields keep track of the flavor and thus allow for flavor off-diagonal transitions.

The dimension-five operators in Eq. (1.14) are best seen as operators of dimension six with the gauge group $SU(2)_L$ taken as the unbroken symmetry of the Lagrangian and the SM fermion grouped, like in the SM, into doublets ψ_L and singlets ψ_R. In this case, the operators contain the Higgs boson field and can be written as

$$\mathcal{L} = \frac{e_D}{2\Lambda^2} \overline{\psi}_L^i \, \sigma_{\mu\nu} \left(\mathbb{D}_M^{ij} + i\gamma_5 \, \mathbb{D}_E^{ij} \right) H \psi_R^j \, F'^{\mu\nu} + \text{H.c.} \tag{1.15}$$

The effective scale is accordingly modulated by the vacuum expectation value (VEV) v_h of the Higgs boson. This VEV keeps track of the chirality breaking, with the whole operator vanishing as v_h goes to zero.

In this review we shall only retain the magnetic dipole $\mathbb{D}_M$ term and set to zero the electric dipole term proportional to $\mathbb{D}_E$. The inclusion of the latter would require the further assumption of CP-odd physics which is, we believe, premature at the moment.

Next, we have the dimension-six operators

$$\mathcal{L}' = \frac{e_D}{2\Lambda^2} \overline{\psi}^i \, \gamma_\mu (\mathbb{R}_r^{ij} + i\gamma_5 \, \mathbb{R}_a^{ij}) D_\nu \psi^j \, F'^{\mu\nu}, \tag{1.16}$$

where the form factor $\mathbb{R}_r$ is related to the charge radius of the fermion; the term $\mathbb{R}_a$ is sometime referred to as the *anapole*.

The operator in Eq. (1.16) contributes, via the equations of motion, to four-fermion operators—which are accounted for in the effective field theory of the dimension-six operators [41] but are not relevant for the massless dark photon interaction to ordinary matter—and to the form factors of the interaction if the particles are off-shell. The latter provide a next-to-leading interaction between the massless dark photon and ordinary matter that has yet to be studied (and is not discussed in this review).

Higher-order operators give vanishingly small contributions and can be neglected.

The scale Λ depends on the parameters of the underlaying UV model. Typically, it is the mass of a heavy state, or the ratio of masses of states of the dark sector, multiplied by the couplings of these states to the SM particles. In particular, the dipole operators in Eq. (1.15), as they require a chirality flip, can turn out to be enhanced, or suppressed, according to the underlaying model chirality mixing.

The fact that the interaction between the massless dark photon and the SM states only takes place through higher-order operators provide an appealing explanation for its weakness. The structure of these operators leads directly to the possible underlaying UV models—a minimal example of which is discussed in Sect. 2.3.

1.3   Dark Matter and the Dark Photon

Dark matter is part of the dark sector. The interplay between the dark photon and dark matter opens new windows on its physics and gives further constraints. Whereas in most scenarios dark matter is one of the fermion (or scalar) states in this sector, there also exists the possibility that dark matter could be a very light vector boson like the massive dark photon itself.

1.3.1   Massless Dark Photon and Galaxy Dynamics

Models of self-interacting dark matter charged under Abelian or non-Abelian gauge groups and interacting through the exchange of massless as well as massive particles have a long history.[1]

The most obvious obstacle to having dark matter in the dark sector interacting via a long-range force as the one carried by the massless dark photon comes from the essentially collisionless dynamics of galaxies and the ellipticity of their dark-matter halo.

The most severe observational limits come from the present dark matter density distribution in collapsed dark matter structures, rather than effects in the early Universe or the early stages of structure formation [48, 49, 59].

Bounds have been derived from the dynamics in merging clusters, such as the Bullet Cluster [65], the tidal disruption of dwarf satellites along their orbits in the host halo, and kinetic energy exchanges among dark matter particles in virialized halos. The latter turns out to be the most constraining bound, noticing that self-interactions tend to isotropize dark matter velocity distributions, while there are galaxies whose gravitational potentials show a triaxial structure with significant velocity anisotropy; limits have been computed, with subsequent refinements, via estimating an isotropization timescale (through hard scattering and cumulative effects of many interactions, also taking into account Debye screening) and comparison to the estimated age of the object [49], or following more closely the evolution of the velocity anisotropy due to the energy transfer [64]. The ellipticity profile inferred for the galaxy NGC720, according to [64] sets a limit of about

$$m_\chi \left(\frac{0.01}{\alpha_{\mathrm{d}}}\right)^{2/3} \gtrsim 300 \text{ GeV} , \qquad (1.17)$$

where m_χ stands for the dark matter mass and the α_{d} scaling quoted is approximate and comes from the leading m_χ over α_{d} scaling in the expression for the isotropization timescale.

---

[1] The literature on the subject is already very extensive, see, for example, [42–64].

Interacting dark matter can form bound states. The phenomenology of such atomic dark matter [51] has been discussed in the literature, see [59] and references therein.

The limit in Eq. (1.17) is subject to a number of uncertainties and assumptions; it is less stringent than earlier results, such as the original bound from soft scattering quoted in [48],

$$\frac{G_N m_\chi^4 N}{8\alpha_D^2} \gtrsim 50 \log \frac{G_N m_\chi^2 N}{2\alpha_D} ,\tag{1.18}$$

where N is the number of dark-matter particles and G_N is Newton's constant, as well about a factor of 3.5 weaker than [49] (see, also, [66, 67]). On the other hand, results on galaxies from N-body simulations in self-interacting dark matter cosmologies [68], which take into account predicted ellipticities and dark matter densities in the central regions, seem to go in the direction of milder constraints, about at the same level or slightly weaker than the value quoted in Eq. (1.17)—again subject to uncertainties, such as the role played by the central baryonic component of NGC720.

1.3.2   Massless Dark Photon and Dark-Matter Relic Density

All the stable fields within the dark sector provide a multicomponent candidate for dark matter whose relic density depends on the value of their couplings to the dark photons and SM fermions (into which they may annihilate, depending on the UV model) and masses.

Not all of the dark fermions contribute to the relic density. If these fermions are relatively light, their dominant annihilation is into dark photons (see Fig. 1.2)

$$\chi \chi \rightarrow A' A' \tag{1.19}$$

with a rate given by

$$\langle \sigma_{\chi\chi \rightarrow A'A'} v \rangle = \frac{2\pi \alpha_d^2}{m_\chi^2} .\tag{1.20}$$

For a strength $\alpha_d \simeq 0.01$, all fermions with masses up to around 1 TeV have a large cross section and their relic density(see Eq. (5.15) in the Sect. 5.2)

$$\Omega_\chi h^2 \approx \frac{2.5 \times 10^{-10} \text{ GeV}^{-2}}{\langle \sigma_{\chi\chi \rightarrow A'A'} v \rangle} \tag{1.21}$$

is only a percent of the critical one; it is roughly 10^{-4} the critical one for dark fermions in the 1 GeV range, even less for lighter states. These dark fermions are not part of dark matter; they have (mostly) converted into dark photons by the time the universe reaches our age and can only be produced in high-energy events. This is fortunate because, as we have seen, they are ruled out as possible dark matter candidates by the limit on galaxy dynamics.

Heavier dark fermions can be dark matter. The dominant annihilation for these is not into dark photons but into SM fermions via the exchange of some messenger field S—the details depending on the underlying UV model—and is proportional to the corresponding coupling which we denote α_L anticipating the discussion in Sect. 2.3—with a thermally averaged cross section approximately given by

$$\langle \sigma_{\chi\chi\to f\bar{f}} v \rangle \simeq \frac{2\pi\alpha_L^2}{m_S^2} \tag{1.22}$$

instead of Eq. (1.20). The critical relic density can be reproduced if, assuming thermal production,

$$2\pi\alpha_L^2 \left(\frac{10\,\text{TeV}}{m_S} \right)^2 \simeq 0.1 . \tag{1.23}$$

These dark matter fermions belonging to the dark sector are in principle detectable through the long range exchange of the massless dark photon and its coupling to the magnetic (o electric) dipole moment of SM matter which is induced at the one loop level in the UV model of the dark sector. The somewhat complementary problem of dark matter having dipole moment and interacting with nuclei through the exchange of a photon has been discussed in [69–75] This dipole interaction is now included within the basis of the operators in the effective field theory of dark matter detection [76–78].

1.3.3 Massive Dark Photon and Light Dark Matter

When dark matter is lighter than the dark photon, and $m_{A'} > 2m_\chi$, the annihilation channel (see Fig. 1.2)

Fig. 1.2 Feynman diagrams for the three processes that are relevant for the discussion of the massive dark photon and dark matter

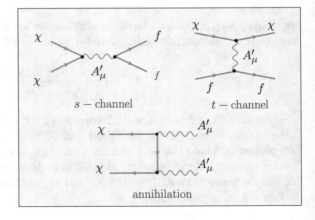

$$\chi\bar{\chi} \to A' \to \bar{f}f \tag{1.24}$$

is open and we are in a scenario with light dark matter (LDM) [79–81]. The cross section is

$$\sigma_{\chi\chi\to ff} = \frac{4\pi}{3}\varepsilon^2\alpha\alpha_d m_\chi^2 \left(1 + \frac{2\,m_e^2}{s}\right)\left(1 + \frac{2\,m_\chi^2}{s}\right)$$

$$\times \frac{s}{(s - m_{A'}^2)^2 + m_{A'}^2\,\Gamma_{A'}^2} \frac{\sqrt{1 - \frac{4\,m_e^2}{m_{A'}^2}}}{\sqrt{1 - \frac{4\,m_\chi^2}{m_{A'}^2}}} \tag{1.25}$$

The thermal average of $\sigma_{\chi\chi\to ff}v$ is defined in Eq. (5.6) of Sect. 5.2. In the non-relativistic limit where $s \simeq 4m_\chi^2$ we have that

$$\langle\sigma_{\chi\chi\to ff}v\rangle \simeq \varepsilon^2\alpha\alpha_d\frac{16\pi m_\chi^2}{(4m_\chi^2 - m_{A'}^2)^2} \tag{1.26}$$

if we neglect m_e with respect to m_χ and $\Gamma_{A'}$ with respect to $m_{A'}$. The thermal average in Eq. (1.26) is related to the relic density ρ_χ (as reviewed in Sect. 5.2), or, in terms of the normalized quantity $\Omega_\chi = \rho_\chi/\rho_c$ as

$$\Omega_\chi\,h^2 \approx \frac{2.5 \times 10^{-10}\ \text{GeV}^{-2}}{\langle\sigma_{\chi\chi\to ff}v\rangle}. \tag{1.27}$$

The general interplay between the massive dark photon and dark matter was originally discussed in [82] and, more recently, in [83].

It has been suggested [84] that the best variable to plot most effectively the constraints in the case of LDM is by means of the *yield* variable

$$y \equiv \varepsilon^2\alpha_d \left(\frac{m_\chi}{m_{A'}}\right)^4 \tag{1.28}$$

because, from Eq. (1.26)

$$\langle\sigma_{\chi\chi\to ff}v\rangle \simeq \frac{16\pi\alpha y}{m_\chi^2} \tag{1.29}$$

and therefore the relic density is brought into the plot. Moreover, the scaling of these limits is made less dependent on the nature of the LDM. We add the limits in the plane $\{y\text{-}m_\chi\}$ to those in the plane $\{\varepsilon\text{-}m_{A'}\}$ in Sect. 3.

The cross section in Eq. (1.25), written in the t-channel (see Fig. 1.2), controls the size of direct detection of dark matter in its scattering off the electrons of the detector thus producing ionization, in particular

$$\sigma_e = \frac{16\pi \mu_{\chi e}^2 \alpha \alpha_d \varepsilon^2}{(m_{A'}^2 + \alpha^2 m_e^2)^2} \left| F(q^2) \right|^2 , \tag{1.30}$$

where $\mu_{\chi e}$ is the reduced mass of the electron and χ, and $F(q^2)$ a form factor given by

$$F(q^2) = \frac{m_{A'}^2 + \alpha^2 m_e^2}{m_{A'}^2 + q^2} , \tag{1.31}$$

with q^2 the square of the exchanged momentum. This relationship translates into a differential event rate in a dark-matter detector with N_T the number of target nuclei per unit mass

$$\frac{dR}{d\ln E} = N_T \frac{\rho_\chi}{m_\chi} \frac{d\langle \sigma_e v \rangle}{d\ln E} , \tag{1.32}$$

where E is the electron energy, $\langle \sigma_e v \rangle$ is the thermally averaged cross section with v the χ velocity, and ρ_χ the local density of χ. This makes possible to utilize limits on LDM direct detection to constraint the dark photon parameter ε [81].

1.3.4 Massive Dark Photon as Dark Matter

A very light massive dark photon could be a dark matter candidate[2] if produced non-thermally in the early Universe as a condensate, the same way as the axion is produced by the *misalignement* mechanism [86–88]. In this mechanism, the value of the field is frozen by the fast expanding Universe to whatever value it has at the initial moment. The rate of expansion is much larger than the mass and the field has no time to relax to the minimum of the potential. The unavoidable (and troublesome) Lorentz-invariance violation is estimated to be small and undetectable.

In this scenario for the dark photon, as discussed in [89, 90], the mass arises via the Stueckelberg mechanism and there must be a non-minimal coupling to gravity. Once the Hubble constant value drops below the mass of the dark photon, its field starts to oscillate and these oscillations behave like non-relativistic matter, that is, like cold dark matter.

There exist two constraints on the parameters of this dark photon scenario. First of all, the initial value must be fine-tuned to reproduce the critical density. Second, the decay into photons and SM leptons must not affect the cosmic microwave background. This latter requirement means that the mixing parameter ε must not be too large (roughly, less than 10^{-9}) and the mass $m_{A'}$ must be less than 1 MeV.

Production by fluctuations during inflation provides another possibility of having a massive dark photon as dark matter [91, 92].

---

[2]In addition, the dark Higgs field breaking the $U(1)$ symmetry can provide yet another dark matter candidate [85].

The dark-photon dark matter is non-relativistic and interacts with ordinary matter mostly through the photo-electric process in which a photon (with energy $m_{A'}$) is captured by an atom, with atomic number Z, with a cross section given, for ordinary photons, by

$$\sigma_{p.e.} = 4\alpha^4 \sqrt{2} Z^5 \frac{8\pi r_e^2}{3} \left(\frac{m_e}{\omega}\right)^{7/2},\tag{1.33}$$

where ω is the photon energy and r_e the classical radius of the electron $r_e = \alpha/m_e$. The cross section for the dark photons is that of ordinary photons rescaled by the mixing parameter ε:

$$\sigma_{A'} = \varepsilon^2 \sigma_{p.e.} .\tag{1.34}$$

This scenario is made accessible to the experiments by considering the rate of absorption of the dark photon by the detector [93, 94]:

$$\Gamma_{A'} = \frac{\rho_{A'}}{m_{A'}} \sigma_{A'} v_{A'},\tag{1.35}$$

where the density $\rho_{A'}$ is estimated from the relic density (or the flux from the Sun).

References

1. B. Holdom, Two U(1)'s and Epsilon charge shifts. Phys. Lett. **166B**, 196–198 (1986). https://doi.org/10.1016/0370-2693(86)91377-8
2. P. Fayet, Extra U(1)'s and new forces. Nucl. Phys. B **347**, 743–768 (1990). https://doi.org/10.1016/0550-3213(90)90381-M
3. P. Fayet, Effects of the Spin 1 partner of the Goldstino (Gravitino) on neutral current phenomenology. Phys. Lett. B **95**, 285–289 (1980). https://doi.org/10.1016/0370-2693(80)90488-8
4. P. Fayet, On the search for a new Spin 1 Boson. Nucl. Phys. B **187**, 184–204 (1981). https://doi.org/10.1016/0550-3213(81)90122-X
5. L.B. Okun, Limits of electrodynamics: paraphotons? Sov. Phys. JETP **56**, 502 (1982). [Zh. Eksp. Teor. Fiz.83,892(1982)]
6. H. Georgi, P.H. Ginsparg, S.L. Glashow, Photon oscillations and the cosmic background radiation. Nature **306**, 765–766 (1983). https://doi.org/10.1038/306765a0
7. J.L. Hewett et al., Fundamental Physics at the Intensity Frontier (2012). arXiv:1205.2671 [hep-ex]. http://lss.fnal.gov/archive/preprint/fermilab-conf-12-879-ppd.shtml. https://doi.org/10.2172/1042577
8. R. Essig et al., Working group report: new light weakly coupled particles, in *Proceedings, 2013 Community Summer Study on the Future of U.S. Particle Physics: Snowmass on the Mississippi (CSS2013): Minneapolis, MN, USA, July 29–Au. 6, 2013* (2013). arXiv:1311.0029 [hep-ph]. http://www.slac.stanford.edu/econf/C1307292/docs/IntensityFrontier/NewLight-17.pdf
9. M. Raggi, V. Kozhuharov, Results and perspectives in dark photon physics. Riv. Nuovo Cim. **38**(10), 449–505 (2015). https://doi.org/10.1393/ncr/i2015-10117-9
10. M.A. Deliyergiyev, Recent progress in search for dark sector signatures. Open Phys. **14**(1), 281–303 (2016). arXiv:1510.06927 [hep-ph]. https://doi.org/10.1515/phys-2016-0034

11. S. Alekhin et al., A facility to search for hidden particles at the CERN SPS: the SHiP physics case. Rept. Prog. Phys. **79**(12), 124201 (2016). arXiv:1504.04855 [hep-ph]. https://doi.org/10.1088/0034-4885/79/12/124201

12. F. Curciarello, Review on dark photon. EPJ Web Conf. **118**, 01008 (2016). https://doi.org/10.1051/epjconf/201611801008

13. J. Alexander et al., Dark Sectors 2016 Workshop: Community Report (2016). arXiv:1608.08632 [hep-ph]. http://lss.fnal.gov/archive/2016/conf/fermilab-conf-16-421.pdf

14. J. Beacham et al., Physics beyond colliders at CERN: beyond the standard model working group report. J. Phys. G **47**(1), 010501 (2020). arXiv:1901.09966 [hep-ex]. https://doi.org/10.1088/1361-6471/ab4cd2

15. B.A. Dobrescu, Massless gauge bosons other than the photon. Phys. Rev. Lett. **94**, 151802 (2005). https://doi.org/10.1103/PhysRevLett.94.151802. arXiv:hep-ph/0411004 [hep-ph]

16. F. del Aguila, M. Masip, M. Perez-Victoria, Physical parameters and renormalization of U(1)-a x U(1)-b models. Physis **B456**, 531–549 (1995). arXiv:hep-ph/9507455 [hep-ph]. https://doi.org/10.1016/0550-3213(95)00511-6

17. D. Feldman, Z. Liu, P. Nath, The Stueckelberg Z-prime extension with kinetic mixing and milli-charged dark matter from the hidden sector. Phys. Rev. D **75**, 115001 (2007). https://doi.org/10.1103/PhysRevD.75.115001. arXiv:hep-ph/0702123 [HEP-PH].

18. S. Davidson, S. Hannestad, G. Raffelt, Updated bounds on millicharged particles. JHEP **05**, 003 (2000). https://doi.org/10.1088/1126-6708/2000/05/003. arXiv:hep-ph/0001179 [hep-ph]

19. H. Ruegg, M. Ruiz-Altaba, The Stueckelberg field. Int. J. Mod. Phys. **A19**, 3265–3348 (2004). arXiv:hep-th/0304245 [hep-th]. https://doi.org/10.1142/S0217751X04019755

20. T. Appelquist, B.A. Dobrescu, A.R. Hopper, Nonexotic neutral Gauge Bosons. Phys. Rev. D **68**, 035012 (2003). https://doi.org/10.1103/PhysRevD.68.035012. arXiv:hep-ph/0212073 [hep-ph]

21. P. Galison, A. Manohar, Two Z's or not two Z's? Phys. Lett. **136B**, 279–283 (1984). https://doi.org/10.1016/0370-2693(84)91161-4

22. X.-G. He, G.C. Joshi, H. Lew, R.R. Volkas, Simplest Z-prime model. Phys. Rev. D **44**, 2118–213 (1991). https://doi.org/10.1103/PhysRevD.44.2118

23. K.S. Babu, C.F. Kolda, J. March-Russell, Implications of generalized Z-Z-prime mixing. Phys. Rev. **D57**, 6788–6792 (1998). arXiv:hep-ph/9710441 [hep-ph]. https://doi.org/10.1103/PhysRevD.57.6788

24. H. Davoudiasl, H.-S. Lee, W.J. Marciano, 'Dark' Z implications for parity violation, rare meson decays, and higgs physics. Phys. Rev. D **85**, 115019 (2012). https://doi.org/10.1103/PhysRevD.85.115019. arXiv:1203.2947 [hep-ph]

25. J. Heeck, Unbroken B? L symmetry. Phys. Lett. **B739**, 256–262 (2014). arXiv:1408.6845 [hep-ph]. https://doi.org/10.1016/j.physletb.2014.10.067

26. M. Bauer, P. Foldenauer, J. Jaeckel, Hunting all the hidden photons. JHEP **07**, 094 (2018). https://doi.org/10.1007/JHEP07(2018)094. arXiv:1803.05466 [hep-ph]. [JHEP18,094(2020)]

27. P. Fayet, The light U boson as the mediator of a new force, coupled to a combination of Q, B, L and dark matter. Eur. Phys. J. **C77**(1), 53 (2017). arXiv:1611.05357 [hep-ph]. https://doi.org/10.1140/epjc/s10052-016-4568-9

28. T.G. Rizzo, Kinetic mixing, dark photons and an extra dimension. Part I. JHEP **07**, 118 (2018). arXiv:1801.08525 [hep-ph]. https://doi.org/10.1007/JHEP07(2018)118

29. E. Bertuzzo, S. Jana, P.A. Machado, R. Zukanovich Funchal, Neutrino masses and mixings dynamically generated by a light dark sector. Phys. Lett. B **791**, 210–214 (2019). arXiv:1808.02500 [hep-ph]. https://doi.org/10.1016/j.physletb.2019.02.023

30. R. Essig, P. Schuster, N. Toro, Probing dark forces and light hidden sectors at low-energy e+e- colliders. Phys. Rev. D **80**, 015003 (2009). https://doi.org/10.1103/PhysRevD.80.015003. arXiv:0903.3941 [hep-ph]

31. S. Koren, R. McGehee, Freezing-in twin dark matter. Phys. Rev. D **101**(5), 055024 (2020). arXiv:1908.03559 [hep-ph]. https://doi.org/10.1103/PhysRevD.101.055024

32. T. Gherghetta, J. Kersten, K. Olive, M. Pospelov, Evaluating the price of tiny kinetic mixing. Phys. Rev. D **100**(9) 095001 (2019). arXiv:1909.00696 [hep-ph]. https://doi.org/10.1103/PhysRevD.100.095001

33. K.R. Dienes, C.F. Kolda, J. March-Russell, Kinetic mixing and the supersymmetric gauge hierarchy. Nucl. Phys. **B492**, 104–118 (1997). arXiv:hep-ph/9610479 [hep-ph]. https://doi. org/10.1016/S0550-3213(97)80028-4, https://doi.org/10.1016/S0550-3213(97)00173-9

34. S.A. Abel, B.W. Schofield, Brane anti-brane kinetic mixing, millicharged particles and SUSY breaking. Nucl. Phys. **B685**, 150–170 (2004). arXiv:hep-th/0311051 [hep-th]. https://doi.org/ 10.1016/j.nuclphysb.2004.02.037

35. S.A. Abel, J. Jaeckel, V.V. Khoze, A. Ringwald, Illuminating the hidden sector of string theory by shining light through a magnetic field. Phys. Lett. **B666**, 66–70 (2008). arXiv:hep-ph/0608248 [hep-ph]. https://doi.org/10.1016/j.physletb.2008.03.076

36. M. Goodsell, Light hidden U(1)s from string theory, in *Proceedings, 5th Patras Workshop on Axions, WIMPs and WISPs (AXION-WIMP 2009): Durham, UK, July 13–17, 2009*, pp. 165–168 (2009). arXiv:0912.4206 [hep-th]. https://doi.org/10.3204/DESY-PROC-2009-05/goodsell_mark

37. M. Goodsell, J. Jaeckel, J. Redondo, A. Ringwald, Naturally light hidden photons in large volume string compactifications. JHEP **11**, 027 (2009). https://doi.org/10.1088/1126-6708/2009/11/027. arXiv:0909.0515 [hep-ph]

38. J.J. Heckman, C. Vafa, An exceptional sector for F-theory GUTs. Phys. Rev. D **83**, 026006 (2011). https://doi.org/10.1103/PhysRevD.83.026006. arXiv:1006.5459 [hep-th]

39. N. Arkani-Hamed, N. Weiner, LHC signals for a superunified theory of dark matter. JHEP **12**, 104 (2008). https://doi.org/10.1088/1126-6708/2008/12/104. arXiv:0810.0714 [hep-ph]

40. Y.F. Chan, M. Low, D.E. Morrissey, A.P. Spray, LHC signatures of a minimal supersymmetric hidden valley. JHEP **05**, 155 (2012). https://doi.org/10.1007/JHEP05(2012)155. arXiv:1112.2705 [hep-ph]

41. B. Grzadkowski, M. Iskrzynski, M. Misiak, J. Rosiek, Dimension-Six terms in the standard model Lagrangian. JHEP **10**, 085 (2010). https://doi.org/10.1007/JHEP10(2010)085. arXiv:1008.4884 [hep-ph]

42. H. Goldberg, and L.J. Hall, A new Candidate for dark matter. Phys. Lett. **B174** 151 (1986). [,467(1986)]. https://doi.org/10.1016/0370-2693(86)90731-8

43. B. Holdom, Searching for ε charges and a new U(1). Phys. Lett. B **178**, 65–70 (1986). https:// doi.org/10.1016/0370-2693(86)90470-3

44. B.-A. Gradwohl, J.A. Frieman, Dark matter, long range forces, and large scale structure. Astrophys. J. **398**, 407–424 (1992). https://doi.org/10.1086/171865

45. E.D. Carlson, M.E. Machacek, L.J. Hall, Self-interacting dark matter. Astrophys. J. **398**, 43–52 (1992). https://doi.org/10.1086/171833

46. R. Foot, Mirror matter-type dark matter. Int. J. Mod. Phys. **D13**, 2161–2192 (2004). arXiv:astro-ph/0407623 [astro-ph]. https://doi.org/10.1142/S0218271804006449

47. J.L. Feng, H. Tu, H.-B. Yu, Thermal relics in hidden sectors. JCAP **0810**, 043 (2008). https:// doi.org/10.1088/1475-7516/2008/10/043. arXiv:0808.2318 [hep-ph]

48. L. Ackerman, M.R. Buckley, S.M. Carroll, M. Kamionkowski, Dark matter and dark radiation. Phys. Rev. D **79**, 023519 (2009). https://doi.org/10.1103/PhysRevD.79.023519,. arXiv:0810.5126 [hep-ph]. [,277(2008)]. https://doi.org/10.1142/9789814293792_0021

49. J.L. Feng, M. Kaplinghat, H. Tu, H.-B. Yu, Hidden charged dark matter. JCAP **0907**, 004 (2009). https://doi.org/10.1088/1475-7516/2009/07/004. arXiv:0905.3039 [hep-ph]

50. N. Arkani-Hamed, D.P. Finkbeiner, T.R. Slatyer, N. Weiner, A theory of dark matter. Phys. Rev. **D79**, 015014 (2009). arXiv:0810.0713 [hep-ph]. https://doi.org/10.1103/PhysRevD.79. 015014

51. D.E. Kaplan, G.Z. Krnjaic, K.R. Rehermann, C.M. Wells, Atomic dark matter. JCAP **1005**, 021 (2010). https://doi.org/10.1088/1475-7516/2010/05/021. arXiv:0909.0753 [hep-ph]

52. M.R. Buckley, P.J. Fox, Dark matter self-interactions and light force carriers. Phys. Rev. D **81**, 083522 (2010). https://doi.org/10.1103/PhysRevD.81.083522. arXiv:0911.3898 [hep-ph]

53. D. Hooper, N. Weiner, W. Xue, Dark forces and light dark matter. Phys. Rev. D **86**, 056009 (2012). https://doi.org/10.1103/PhysRevD.86.056009. arXiv:1206.2929 [hep-ph]

54. L.G. van den Aarssen, T. Bringmann, C. Pfrommer, Is dark matter with long-range interactions a solution to all small-scale problems of Λ CDM cosmology? Phys. Rev. Lett. **109**, 231301 (2012). https://doi.org/10.1103/PhysRevLett.109.231301. arXiv:1205.5809 [astro-ph.CO]

55. J.M. Cline, Z. Liu, W. Xue, Millicharged atomic dark matter. Phys. Rev. D **85**, 101302 (2012). https://doi.org/10.1103/PhysRevD.85.101302. arXiv:1201.4858 [hep-ph]
56. S. Tulin, H.-B. Yu, K.M. Zurek, Beyond Collisionless dark matter: particle physics dynamics for dark matter halo structure. Phys. Rev. **D87**(11), 115007 (2013). arXiv:1302.3898 [hep-ph]. https://doi.org/10.1103/PhysRevD.87.115007
57. E. Gabrielli, M. Raidal, Exponentially spread dynamical Yukawa couplings from nonperturbative chiral symmetry breaking in the dark sector. Phys. Rev. **D89**(1), 015008 (2014).. arXiv:1310.1090 [hep-ph]. https://doi.org/10.1103/PhysRevD.89.015008
58. M. Baldi, Structure formation in multiple dark matter cosmologies with long-range scalar interactions. Mon. Not. Roy. Astron. Soc. **428**, 2074 (2013). https://doi.org/10.1093/mnras/sts169. arXiv:1206.2348 [astro-ph.CO]
59. F.-Y. Cyr-Racine, K. Sigurdson, Cosmology of atomic dark matter. Phys. Rev. **D87**(10), 103515 (2013). arXiv:1209.5752 [astro-ph.CO]. https://doi.org/10.1103/PhysRevD.87.103515
60. J.M. Cline, Z. Liu, G. Moore, W. Xue, Composite strongly interacting dark matter. Phys. Rev. **D90**(1), 015023 (2014). arXiv:1312.3325 [hep-ph]. https://doi.org/10.1103/PhysRevD.90.015023
61. X. Chu, B. Dasgupta, Dark radiation alleviates problems with dark matter Halos. Phys. Rev. Lett. **113**(16), 161301 (2014). arXiv:1404.6127 [hep-ph]. https://doi.org/10.1103/PhysRevLett.113.161301
62. K.K. Boddy, J.L. Feng, M. Kaplinghat, T.M.P. Tait, Self-Interacting dark matter from a nonabelian hidden sector. Phys. Rev. **D89**(11), 115017 (2014). arXiv:1402.3629 [hep-ph]. https://doi.org/10.1103/PhysRevD.89.115017
63. M.A. Buen-Abad, G. Marques-Tavares, M. Schmaltz, Non-Abelian dark matter and dark radiation. Phys. Rev. **D92**(2), 023531 (2015). arXiv:1505.03542 [hep-ph]. https://doi.org/10.1103/PhysRevD.92.023531
64. P. Agrawal, F.-Y. Cyr-Racine, L. Randall, J. Scholtz, Make dark matter charged again. JCAP **1705**(05), 022 (2017). arXiv:1610.04611 [hep-ph]. https://doi.org/10.1088/1475-7516/2017/05/022
65. D. Clowe, M. Bradac, A.H. Gonzalez, M. Markevitch, S.W. Randall, C. Jones, D. Zaritsky, A direct empirical proof of the existence of dark matter. Astrophys. J. **648**, L109–L113 (2006). arXiv:astro-ph/0608407 [astro-ph]. https://doi.org/10.1086/508162
66. J.L. Feng, M. Kaplinghat, H.-B. Yu, Halo shape and relic density exclusions of sommerfeld-enhanced dark matter explanations of Cosmic Ray excesses. Phys. Rev. Lett. **104**, 151301 (2010). https://doi.org/10.1103/PhysRevLett.104.151301. arXiv:0911.0422 [hep-ph]
67. T. Lin, H.-B. Yu, K.M. Zurek, On symmetric and asymmetric light dark matter. Phys. Rev. D **85**, 063503 (2012). https://doi.org/10.1103/PhysRevD.85.063503. arXiv:1111.0293 [hep-ph]
68. A.H.G. Peter, M. Rocha, J.S. Bullock, M. Kaplinghat, Cosmological simulations with self-interacting dark matter II: halo shapes versus observations. Mon. Not. Roy. Astron. Soc. **430**, 105 (2013). arXiv:1208.3026 [astro-ph.CO]. https://doi.org/10.1093/mnras/sts535
69. M. Pospelov, T. ter Veldhuis, Direct and indirect limits on the electromagnetic form-factors of WIMPs. Phys. Lett. B **480**, 181–186 (2000). arXiv:hep-ph/0003010. https://doi.org/10.1016/S0370-2693(00)00358-0
70. K. Sigurdson, M. Doran, A. Kurylov, R.R. Caldwell, M. Kamionkowski, Dark-matter electric and magnetic dipole moments. Phys. Rev. **D70**, 083501 (2004). arXiv:astro-ph/0406355 [astro-ph]. [Erratum: Phys. Rev.D73,089903(2006)]. https://doi.org/10.1103/PhysRevD.70.083501
71. T. Banks, J.-F. Fortin, S. Thomas, Direct Detection of Dark Matter Electromagnetic Dipole Moments. arXiv:1007.5515 [hep-ph]
72. V. Barger, W.-Y. Keung, D. Marfatia, Electromagnetic properties of dark matter: Dipole moments and charge form factor. Phys. Lett. B **696**, 74–78 (2011). arXiv:1007.4345 [hep-ph]. https://doi.org/10.1016/j.physletb.2010.12.008
73. N. Fornengo, P. Panci, M. Regis, Long-Range forces in direct dark matter searches. Phys. Rev. D **84**, 115002 (2011). https://doi.org/10.1103/PhysRevD.84.115002. arXiv:1108.4661 [hep-ph]
74. E. Del Nobile, C. Kouvaris, P. Panci, F. Sannino, J. Virkajarvi, Light magnetic dark matter in direct detection searches. JCAP **08**, 010 (2012). https://doi.org/10.1088/1475-7516/2012/08/010. arXiv:1203.6652 [hep-ph]

75. X. Chu, J.-L. Kuo, J. Pradler, Dark sector-photon interactions in proton-beam experiments. Phys. Rev. D **101**, 075035 (2020). https://doi.org/10.1103/PhysRevD.101.075035. arXiv:2001.06042 [hep-ph]

76. A.L. Fitzpatrick, W. Haxton, E. Katz, N. Lubbers, Y. Xu, The effective field theory of dark matter direct detection. JCAP **1302**, 004 (2013). https://doi.org/10.1088/1475-7516/2013/02/004. arXiv:1203.3542 [hep-ph]

77. S. Liem, G. Bertone, F. Calore, R. Ruiz de Austri, T.M.P. Tait, R. Trotta, C. Weniger, Effective field theory of dark matter: a global analysis. JHEP **09**, 077 (2016). https://doi.org/10.1007/JHEP09(2016)077. arXiv:1603.05994 [hep-ph]

78. J. Brod, A. Gootjes-Dreesbach, M. Tammaro, J. Zupan, Effective field theory for dark matter direct detection up to Dimension Seven. JHEP **10**, 065 (2018). https://doi.org/10.1007/JHEP10(2018)065. arXiv:1710.10218 [hep-ph]

79. C. Boehm, T. Ensslin, J. Silk, Can Annihilating dark matter be lighter than a few GeVs J. Phys. G **30**, 279–286 (2004). arXiv:astro-ph/0208458. https://doi.org/10.1088/0954-3899/30/3/004

80. S. Knapen, T. Lin, K.M. Zurek, Light dark matter: models and constraints. Phys. Rev. **D96**(11), 115021 (2017). arXiv:1709.07882 [hep-ph]. https://doi.org/10.1103/PhysRevD.96.115021

81. R. Essig, J. Mardon, T. Volansky, Direct detection of Sub-GeV dark matter. Phys. Rev. D **85**, 076007 (2012). https://doi.org/10.1103/PhysRevD.85.076007. arXiv:1108.5383 [hep-ph]

82. C. Boehm, P. Fayet, Scalar dark matter candidates. Nucl. Phys. B **683**, 219–263 (2004). arXiv:hep-ph/0305261. https://doi.org/10.1016/j.nuclphysb.2004.01.015

83. T. Hambye, M.H. Tytgat, J. Vandecasteele, L. Vanderheyden, Dark matter from dark photons: a taxonomy of dark matter production. Phys. Rev. D **100**(9), 095018 (2019). arXiv:1908.09864 [hep-ph]. https://doi.org/10.1103/PhysRevD.100.095018

84. E. Izaguirre, G. Krnjaic, P. Schuster, N. Toro, Analyzing the discovery potential for light dark matter. Phys. Rev. Lett. **115**(25), 251301 (2015). arXiv:1505.00011 [hep-ph]. https://doi.org/10.1103/PhysRevLett.115.251301

85. C. Mondino, M. Pospelov, J.T. Ruderman, O. Slone, Dark Higgs Dark Matter. arXiv:2005.02397 [hep-ph]

86. J. Preskill, M.B. Wise, F. Wilczek, Cosmology of the invisible axion. Phys. Lett. **120B**, 127–132 (1983). https://doi.org/10.1016/0370-2693(83)90637-8

87. L.F. Abbott, P. Sikivie, A cosmological bound on the invisible axion. Phys. Lett. **120B**, 133–136 (1983). https://doi.org/10.1016/0370-2693(83)90638-X

88. M. Dine, W. Fischler, The not so harmless axion. Phys. Lett. **120B**, 137–141 (1983). https://doi.org/10.1016/0370-2693(83)90639-1

89. A.E. Nelson, J. Scholtz, Dark light, dark matter and the misalignment mechanism. Phys. Rev. D **84**, 103501 (2011). https://doi.org/10.1103/PhysRevD.84.103501. arXiv:1105.2812 [hep-ph]

90. P. Arias, D. Cadamuro, M. Goodsell, J. Jaeckel, J. Redondo, A. Ringwald, WISPy cold dark matter. JCAP **1206**, 013 (2012). https://doi.org/10.1088/1475-7516/2012/06/013. arXiv:1201.5902 [hep-ph]

91. P.W. Graham, J. Mardon, S. Rajendran, Vector dark matter from inflationary fluctuations. Phys. Rev. D **93**(10), 103520 (2016). arXiv:1504.02102 [hep-ph]. https://doi.org/10.1103/PhysRevD.93.103520

92. Y. Nakai, R. Namba, Z. Wang, Light Dark Photon Dark Matter from Inflation. arXiv:2004.10743 [hep-ph]

93. M. Pospelov, A. Ritz, M.B. Voloshin, Bosonic super-WIMPs as keV-scale dark matter. Phys. Rev. D **78**, 115012 (2008). https://doi.org/10.1103/PhysRevD.78.115012. arXiv:0807.3279 [hep-ph]

94. I.M. Bloch, R. Essig, K. Tobioka, T. Volansky, T.-T. Yu, Searching for dark absorption with direct detection experiments. JHEP **06**, 087 (2017). https://doi.org/10.1007/JHEP06(2017)087. arXiv:1608.02123 [hep-ph]

Chapter 2
Phenomenology of the Massless Dark Photon

The phenomenology of the massless dark photon depends on the effect of the higher-order operator in Eq. (1.15) which mediates its interaction with the SM particles. This operator enters the measured observables with an effective scale Λ and the absolute value

$$d_M^{ij} \equiv |\mathbb{D}_M^{ij}| \tag{2.1}$$

of the magnetic dipole coefficient (neglecting the CP-odd $\mathbb{D}_E$) which can eventually be related to the parameters of the underlying UV model like masses and coupling constants. The experimental searches can thus be framed in terms of the scale Λ, the dipole coefficient d_M^{ij} and the dark charge coupling e_D, which we rewrite as $\alpha_D = e_D^2/4\pi$. We do not assume this scale and coefficient to be universal. Depending on the particular experimental set-up, the constraints are further sensitive to which particular lepton or quark is actually taking part in the interaction. The index, or indices, i and j keep track on the flavor dependence.

We discuss in Sect. 2.2 the other side of the massless dark photon, namely the search for dark particles coupled to the ordinary photon by a milli-charge.

2.1 Limits on the Dark Dipole Scale d_M/Λ^2

We collect in this section the known constraints on the size of the operator in Eq. (1.15) (Fig. 2.3).

We show in Figs. 2.1 and 2.2 the more stringent limits. Though these limits are on the combinations d_M/Λ^2, with a factor depending on α_D, we find it convenient to plot them as d_M as a function of Λ so as to easily see what values of the dipole coefficient are allowed given a value for the scale Λ (and two representative value of α_D).

© The Author(s), under exclusive license to Springer Nature Switzerland AG 2021
M. Fabbrichesi et al., *The Physics of the Dark Photon*,
SpringerBriefs in Physics, https://doi.org/10.1007/978-3-030-62519-1_2

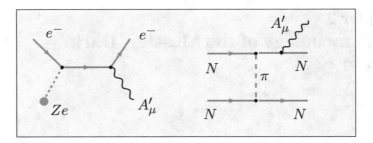

Fig. 2.1 *Bremsstrahlung* of dark photons from electrons in a star and from nucleons in a supernova

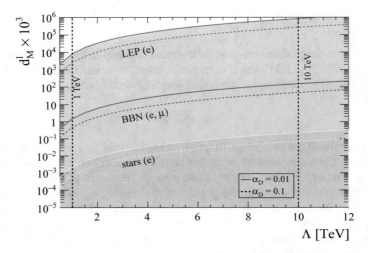

Fig. 2.2 Model-independent limits for the interaction with **leptons**. The limits on the dark dipole operator d_M^ℓ/Λ^2 are shown by taking the coefficient d_M^ℓ as a function of the scale Λ (for two representative values of α_D). Given an energy scale, the allowed values for d_M^ℓ can be read from the plot. The strongest bound on electrons comes from stellar cooling (stars). Big bang nucleosynthesis (BBN) and collider physics (LEP) set the other depicted bounds. Solid lines are for the representative value $\alpha_D = 0.01$, dashed lines for $\alpha_D = 0.1$

2.1.1   Astrophysics and Cosmology

Astrophysics and cosmology provide very stringent limits on the interaction of the dark photon with SM matter as given by the operator in Eq. (1.15). It is understood that all the limits are mostly on the order of magnitude because of intrinsic uncertainties in the astrophysics of stellar medium, supernova dynamics and cosmological processes.

Astrophysical constraints for models with a massless dark photon can be derived from those obtained for axion-like particles because the dipole operator in Eq. (1.15) gives, in the non-relativistic limit, a derivative (and spin-dependent) coupling of the dark photon with momentum **k** and polarization ϵ to ordinary fermions ψ given by

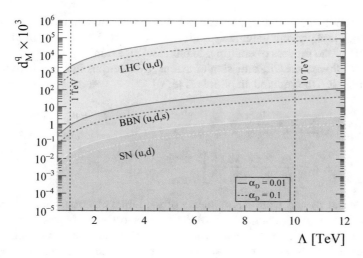

Fig. 2.3 Model-independent limits for for the interaction with **quarks**. Same as in Fig. 2.2. The strongest bounds on light quarks comes from supernovae (SN). Primordial nucleosynthesis (BBN) and collider physics (LHC) set the other depicted bounds. Solid lines are for the representative value $\alpha_D = 0.01$, dashed lines for $\alpha_D = 0.1$

$$\mathcal{M}_{A'_\mu} \approx \bar\psi \, (\mathbf{k} \times \boldsymbol{\epsilon}) \cdot \boldsymbol{\sigma} \, \psi \,, \tag{2.2}$$

which, after averaging over the polarizations, gives the same contribution as that for a pseudo-scalar particle [1, 2] like the axion, namely

$$\mathcal{M}_a \approx \bar\psi \, \mathbf{k} \cdot \boldsymbol{\sigma} \, \psi \,. \tag{2.3}$$

Only a factor of two must be included for the independent polarizations of the dark photon.

Because the massless dark photon does not mix with the ordinary photon, we can compute the limits in a kinetic theory in which the amplitude for the relevant process is computed in the vacuum and the effect of the medium—be it the stellar interior or the supernova nucleon gas—is included in the abundances of the SM states at the given temperature.

• Stars. The luminosity of stars is related to their energy balance. This balance is a sensitive probe of the stellar dynamics and the particle-physics processes on which is based. Three processes are important for energy loss in stars: Compton scattering, pair creation and *Bremsstrahlung* (Fig. 2.1). Of these three, it is the latter that provides the most stringent limit. The non-observation of anomalous energy transport, in various different types of stars, places strong constraints on the dipole coupling between SM states and the dark photon [3, 4].

The quantity we need is the energy loss due to the emission of the extra particle. The energy loss per unit volume Q is given in the Sect. 5.3 in terms of the squared amplitude of the process of emitting, in our case, an axion.

For the *Bremsstrahlung* emission of axions, by electrons in the field of n_j nuclei with charge Z_j, the squared amplitude is [5, 6]

$$\sum_{\text{spin}} |\mathcal{M}|^2 = \sum_j Z_j^2 n_j \frac{4\alpha^2 \alpha'_{ae}}{\pi} \frac{|\mathbf{p}_1||\mathbf{p}_2|\omega^2}{(\mathbf{q}^2 + \kappa_F^2)^2} \left[2\omega^2 \frac{p_1 \cdot p_2 - m_e^2 + (p_2 - p_1) \cdot k}{(p_1 \cdot k)(p_2 \cdot k)} \right.$$
$$\left. + 2 - \frac{p_1 \cdot k}{p_2 \cdot k} - \frac{p_2 \cdot k}{p_1 \cdot k} \right] \tag{2.4}$$

where p_1 and p_2 and k are the momenta of the initial electrons and $q = p_2 - p_1$. ω and k the energy and momentum of the axion and $\kappa_F = (4\alpha p_F E_F/\pi)^{1/2}$ where p_F and E_F are the Fermi momentum and energy of the electrons in the plasma. The coefficient α'_{ae} is the coupling constant of the axion to the electrons.

In a degenerate medium (like the one for red giants and white dwarves) we have that the energy-loss rate per unit mass Q/ρ is given by [2]

$$Q/\rho = \frac{\pi^2 \alpha^2 \alpha'_{ae}}{15} \frac{T^4}{m_e^2} \sum_j Z_j^2 n_j \, F(\kappa_F)$$
$$\simeq \alpha'_{ae} \, 1.08 \times 10^{27} \left(\frac{T}{10^8 \text{K}} \right)^4 \frac{Z^2}{A} F(\kappa_F), \tag{2.5}$$

the latter equation is written in units of erg g^{-1}s^{-1}, and the factor F is approximately given in the relativistic limit as

$$F(\kappa_F) \simeq \frac{2 + \kappa_F^2}{2} \ln \frac{2 + \kappa_F^2}{\kappa_F^2} - 1. \tag{2.6}$$

The most stringent limit for electrons comes from cooling in white dwarves [7] and giant red stars [8] by axion *Bremsstrahlung* in a degenerate medium. A combined fit of the data [9] finds (at 2σ) that the coupling must be

$$\alpha'_{ae} \le 3.0 \times 10^{-27}. \tag{2.7}$$

The bound in Eq. (2.7) is translated into a bound for the dark photon by identifying the combination of parameters in the operator in Eq. (1.15) that controls the same process. This correspondence yields the equation

$$\alpha'_{ae} = 2\frac{1}{4\pi} \left(2 e_D d_M^e \frac{v_h m_e}{\Lambda^2} \right)^2, \tag{2.8}$$

where the factor of 2 in front takes into account the two polarizations of the dark photon (with respect to the axion), $v_h = 174$ GeV and m_e is the electron's mass.

To satisfy the limit in Eq. (2.7), the dark photon parameters in Eq. (2.8) must satisfy

$$\frac{\Lambda^2}{\sqrt{\alpha_D}\, d_M^e} \gtrsim 4.5 \times 10^6 \text{ TeV}^2 , \qquad (2.9)$$

after having included the numerical values of m_e and v_h. The limit in Eq. (2.9) updates the one found in [4].

• Supernovae. An additional limit is found from the neutrino signal of supernova 1987A, for which the length of the burst constrains anomalous energy losses in the explosion.

As before, a bound can be derived from that for the coupling between axions and nucleons. The corresponding averaged square amplitude is given in [10, 11] as

$$\sum_{\text{spin}} |\mathcal{M}|^2 = \frac{16(4\pi)^3 \alpha_\pi^2 \, \alpha'_{aN}}{3\, m_N^2} \left[\left(\frac{\mathbf{k}^2}{\mathbf{k}^2 + m_\pi^2} \right)^2 + \left(\frac{\mathbf{l}^2}{\mathbf{l}^2 + m_\pi^2} \right)^2 \right.$$
$$\left. + \frac{\mathbf{k}^2 \mathbf{l}^2 - 3(\mathbf{k} \cdot \mathbf{l})^2}{(\mathbf{k}^2 + m_\pi^2)(\mathbf{l}^2 + m_\pi^2)} \right] \qquad (2.10)$$

where $\alpha_\pi = (2m_N f/m_\pi)^2/4\pi \simeq 15$ is the pion-nucleon coupling and $\mathbf{k} = \mathbf{p}_2 - \mathbf{p}_4$ and $\mathbf{l} = \mathbf{p}_2 - \mathbf{p}_3$ and $\mathbf{p}_i$ the momenta of the nucleons. The coefficient α'_{aN} is the coupling constant of the axion to the nucleons.

In the thermal medium $\mathbf{k}^2 \simeq 3m_N T$ and we can neglect the pion mass to obtain

$$\sum_{\text{spin}} |\mathcal{M}|^2 = \frac{32(4\pi)^3 \alpha_\pi^2 \, \alpha'_{aN}}{m_N^2} \qquad (2.11)$$

and the energy-loss rate per unit mass in the degenerate case is [10, 12]

$$\mathcal{Q}/\rho \simeq \alpha'_{aN}\, 1.74 \times 10^{33} \frac{\rho}{10^{15}} \left(\frac{T}{\text{MeV}} \right)^6 , \qquad (2.12)$$

in units of erg g^{-1}s^{-1}, which should not exceed the neutrino luminosity. This limit yields, taking the most conservative estimate in [13, 14],

$$\alpha'_{aN} \le 1.3 \times 10^{-18} . \qquad (2.13)$$

The combination that controls energy transfer to dark photons in this process from ordinary matter (the quarks in the nucleons) is

$$\alpha'_{aN} = 2 \frac{1}{4\pi} \left(2 e_D d_M^q \frac{v_h\, m_N}{\Lambda^2} \right)^2 , \qquad (2.14)$$

where m_N is the nucleon mass. By taking the limit in Eq. (2.13), we have

$$\frac{\Lambda^2}{\sqrt{\alpha_D}\, d_M^q} \gtrsim 4.3 \times 10^5 \text{ TeV}^2 \,, \tag{2.15}$$

which applies to the light u and d quarks—if we neglect small corrections due to the form factors in going from the nucleons to the quarks. The limit in Eq. (2.15) updates the one found in [4].

A caveat in the limit in Eq. (2.13) is due to the fact that if the coupling is too strong the emitted axions are re-absorbed by the expanding supernova and there is no energy loss; this happens for

$$\alpha'_{aN} \geq 0.7 \times 10^{-14} \,, \tag{2.16}$$

which yields

$$\frac{\Lambda^2}{\sqrt{\alpha_D} d_M^q} \lesssim 5.9 \times 10^3 \text{ TeV}^2 \,, \tag{2.17}$$

There are however limits from laboratory physics, discussed in the next section, that almost close this window.

• Big bang nucleosynthesis. A cosmological bound for the dark photon operator in Eq. (1.15) comes from the determination of the effective number of relativistic species in addition to those of the SM partaking in the thermal bath—the same way the number of neutrinos is constrained. This number is constrained by data on big bang nucleosynthesis (BBN) to be [15]:

$$N_{\text{eff}} = 2.878 \pm 0.278 \,. \tag{2.18}$$

We follow [4] in deriving the corresponding limits.

The two degrees of freedom of the dark photon exceeds this limit at the big bang temperature T_{BBN} and must have decoupled before at temperature T_d which is taken to be just above the QCD phase transition: $T_d = 150$ MeV. The request of decoupling before the BBN epoch can be translated in having the Hubble constant (see Sect. 5.2)

$$H(T_d) = \frac{T_d^2}{M_{Pl}} \left(\frac{\pi^2}{90} g_*(T_d) \right)^{1/2} \tag{2.19}$$

be larger than the rate of interactions between SM states and the dark photon

$$\Gamma_{A'} = n_{A'} \langle \sigma v \rangle \,, \tag{2.20}$$

where $\langle \sigma v \rangle$ is the thermally averaged cross section for the interaction of the dark photon with the SM particles present at the temperature T_d, $v = 1$, and the number density of dark photon is given (see Sect. 5.2) by ($k_B = 1$)

$$n_{A'} = \frac{2\zeta(3)}{\pi^2} T^3 .$$ (2.21)

The cross section for SM fermions to Compton and annihilate into dark photon is approximately given by

$$\langle \sigma v \rangle \simeq \frac{\alpha_D d_M^2 v_h^2}{\Lambda^4} .$$ (2.22)

We thus find the condition

$$\frac{2\zeta(3)}{\pi^2} T_d^3 \langle \sigma v \rangle < \frac{T_d^2}{M_{Pl}} \left(\frac{2\pi^2}{45} g_*(T_d) \right)^{1/2} ,$$ (2.23)

where the effective number of degrees of freedom $g_*(T_d)$ is bound from the limit on N_{eff}. This relationship is obtained from

$$\left(\frac{T_{BBN}}{T_d} \right)^4 = \left(\frac{g_*(T_{BBN})}{g_*(T_d)} \right)^{4/3} < \frac{7}{4} \Delta N_{\text{eff}} ,$$ (2.24)

which, knowing that $g_*(T_{BBN}) = 43/4$, gives

$$g_*(T_d) > (43/7)^{4/3} \Delta N_{\text{eff}}^{-3/4} ,$$ (2.25)

where $\Delta N_{\text{eff}} \equiv N_{\text{eff}} - 3 \simeq 0.34$ by taking 2σ of the result in Eq. (2.18).

The limit applies to the interaction of leptons (electron and muon):

$$\frac{\Lambda^2}{\sqrt{\alpha_D}\, d_M^\ell} \geq 6.6 \times 10^3 \text{ TeV}^2 ,$$ (2.26)

and quarks (s, u, d):

$$\frac{\Lambda^2}{\sqrt{\alpha_D}\, d_M^q} \geq 4.3 \times 10^3 \text{ TeV}^2 ,$$ (2.27)

which partake into the Compton and annihilation processes. The difference between Eqs. (2.26) and (2.27) is due to the number of colors.

2.1.2 Precision, Laboratory and Collider Physics

Laboratory physics can set new constrains on the dipole operator in Eq. (1.15). They are less stringent than those from astrophysics and cosmology because the higher-order dipole operator always yields a small number of events; these small numbers are amplified in the stars by the enormous density of particles in the medium but not in the laboratory experiments where the density is smaller.

• Precision physics. The operator in Eq. (1.15) gives rise to a macroscopic spin-dependent (non-relativistic) potential [16]:

$$V(r) = -\frac{\alpha_D v^2 d_M^a d_M^b}{4 \Lambda^4 r^3} \left[\sigma_a \cdot \sigma_b - 3 (\sigma_a \cdot \hat{r})(\sigma_b \cdot \hat{r}) \right], \tag{2.28}$$

where $\mathbf{r} = \mathbf{r_a} - \mathbf{r_b}$ is the vector distance and $r = |\mathbf{r}|$ and $\hat{r}$ the corresponding unit vector. The potential in Eq. (2.28) is between two fermions f_a and f_b, with spin σ_a and σ_b, and magnetic dipole moments $d_M^{a,b}$, as defined in Eq. (1.15)—whose interaction can affect atomic energy levels as well as macroscopic forces.

The potential in Eq. (2.28) can be used to explore atomic physics as well as macroscopic fifth-force like interactions.

Many atomic energy levels are known with high precision. Unfortunately, the theoretical computation is lagging behind many of the experiments, mainly because of uncertainties in higher-order corrections like those due to the size of the nuclei. For this reason many results are given as energy differences where corrections proportional to $1/r^3$ are factorized out. This procedure makes often impossible to use these results to test the potential in Eq. (2.28).

The best limit is obtained in the fine-structure spectroscopy of Helium. The extra interaction between the two electrons has been discussed in [17] whose limits, obtained by the constraints from the $2^3 P_2$-$2^3 P_1$ transitions in He, can be expressed as

$$\frac{\Lambda^2}{\sqrt{\alpha_D} d_M^e} \gtrsim 872 \text{ GeV}^2 . \tag{2.29}$$

Bounds on long-range forces depending on spin set limits on the scale of the operator in Eq. (1.14) based on the potential in Eq. (2.28) as discussed in [16]. The strongest bounds come from limits on macroscopic forces between electrons [18]

$$\frac{\Lambda^2}{\sqrt{\alpha_D} d_M^e} \gtrsim 1.61 \text{ TeV}^2 , \tag{2.30}$$

and electrons and nucleons [19]

$$\frac{\Lambda^2}{\sqrt{\alpha_D} \sqrt{d_M^e d_M^q}} \gtrsim 1.94 \text{ TeV}^2 . \tag{2.31}$$

The limits among nucleons and electrons and protons are weaker.

Whereas the strong limits on the anomalous magnetic moments of the electron and the muon are traditionally used to set limits on new physics, they cannot be used directly in our case because they only apply to operators coupling to the visible photon. The operator in Eq. (1.14) enters in the computation of the magnet moments but only at higher order with two insertions in the loop computation. The limits are accordingly weak. The contribution of the dark photon to the anomalous magnetic moment is given by

$$a_\ell^{A'} = -\frac{3}{2}\frac{\alpha_D}{\pi}\left(\frac{m_\ell v_h d_M^\ell}{\Lambda^2}\right)^2\left[\frac{5}{4}+\log\frac{\mu^2}{m^2}\right] \tag{2.32}$$

in the $\overline{MS}$ scheme; contrary to the SM case, the result depends on the subtraction of a divergence.

We discuss below in Sect. 2.3 how in the UV model, where there are states coupled to both the dark and the visible photon, the anomalous magnetic moment can be brought to bear directly on the limits.

The quantity Δa_e, the difference between the experimental value of the electron anomalous magnetic moment [20] and its SM prediction is very small. The uncertainty on this difference (at 1σ) is given by [21]

$$\delta_{\Delta a_e} < 8.1 \times 10^{-13} . \tag{2.33}$$

By requiring that the contribution of the dark photon does not exceed this value, and therefore does not contribute to the electron magnetic moment, we obtain

$$\frac{\Lambda^2}{\sqrt{\alpha_D}d_M^e} \gtrsim 0.075 \text{ TeV}^2 , \tag{2.34}$$

by taking the renormalization scale $\mu = m_e$.

The analogous quantity Δa_μ, the difference between the experimental value of the muon anomalous magnetic moment [22] and its SM prediction [23], is less than

$$\Delta a_\mu < 2.74 \times 10^{-9} , \tag{2.35}$$

at 2σ level, from which we derive

$$\frac{\Lambda^2}{\sqrt{\alpha_D}d_M^\mu} \gtrsim 0.5 \text{ TeV}^2 , \tag{2.36}$$

for $\mu = m_\mu$. Notice that the current 3.2σ discrepancy in Δa_μ could be explained by requiring

$$\frac{\Lambda^2}{\sqrt{\alpha_D}d_M^\mu} \simeq 0.27 \text{ TeV}^2 . \tag{2.37}$$

Flavor changing processes can provide constraints on possible dipole operator contributions to off-diagonal interactions.

In the lepton sector the process $\mu \to eX^0$, with X^0 a massless neutral boson, is bounded to [24]

$$\text{BR}\,(\mu \to eX^0) < 5.8 \times 10^{-5}\,, \tag{2.38}$$

which gives

$$\frac{\Lambda^2}{\sqrt{\alpha_D} d_M^{\mu e}} \gtrsim 5.1 \times 10^5 \text{ TeV}^2\,. \tag{2.39}$$

Similar limits in the hadron sector on, for example, the decays $K \to \pi X^0$ or $B \to K X^0$, cannot be used because they are forbidden when X^0 is a spin one boson like the dark photon. The decay $B \to K^* X^0$ is not forbidden but gives a very weak bound. Instead, the current limit on the rare decay $K^+ \to \pi^+ \nu \bar{\nu}$ given by (at the 90% CL, see, for example, [25])

$$\text{BR}\,(K^+ \to \pi^+ \nu \bar{\nu}) < 1.85 \times 10^{-10} \tag{2.40}$$

can be used, if we assume the dark photon to decay into light dark-sector fermions, and yields

$$\frac{\Lambda^2}{\sqrt{\alpha_D} d_M^{sd}} \gtrsim 9.5 \times 10^6 \text{ TeV}^2\,, \tag{2.41}$$

which is the strongest among all the limits on the dipole interaction we have discussed.

• Laboratory physics. An interesting limit is derived by means, again, of the data from SN 1987A, this time indirectly from the counting of events in the Kamiokande detector. Axions from the star can, via inverse *Bremsstrahlung*, excite the oxygen nuclei in the water tank as, in the process $a^{16}O \to {}^{16}O^*$, which subsequently decay producing γ rays triggering the detector. The failure of observing these extra events excludes the values for the coupling α'_{aN} [26]

$$6.5 \times 10^{-14} \le \alpha'_{aN} \le 8.0 \times 10^{-8}\,, \tag{2.42}$$

which can be turned, taking the lower limit in Eq. (2.42), in

$$\frac{\Lambda^2}{\sqrt{\alpha_D} d_M^q} \gtrsim 1.9 \times 10^3 \text{ TeV}^2\,, \tag{2.43}$$

for the massless dark photons. The limit in Eq. (2.43) nicely closes the range left open by Eq. (2.15). A thin windows between Eqs. (2.16) and (2.42) is apparently left open for $\alpha'_{aN} \simeq 10^{-14}$.

• Collider physics. Limits from colliders are weaker but are worthwhile to be reported since they come from laboratory physics which is independent of all astrophysical assumptions. The process of pair annihilation into a dark and an ordinary photon provides a striking benchmark (mono-photon plus missing energy) for this search. It applies to electrons in searches at the LEP [27–29]:

$$\frac{\Lambda^2}{\sqrt{\alpha_D}d_M^e} \gtrsim 1.2 \text{ TeV}^2\,, \qquad\qquad (2.44)$$

and the first generation of quarks at the LHC from CMS [30] with luminosity of 35.9 fb^{-1} (the ATLAS result [31] is with smaller luminosity and less stringent):

$$\frac{\Lambda^2}{\sqrt{\alpha_D}d_M^q} \gtrsim 4.3 \text{ TeV}^2\,. \qquad\qquad (2.45)$$

We computed the limits in Eqs. (2.44) and (2.45) for this review by requiring that the number of dark photon events be, bin by bin, less than the difference between the observed and the expected number of events.

2.1.3 Can the Massless Dark Photon Be Seen at All?

The limits for the dark dipole of the massless dark photon, as summarized in Figs. 2.2 and 2.3, are indeed very stringent. For an effective scale Λ around 1 TeV, for example, only values of dipole moments of $O(10^{-6})$ for electrons and $O(10^{-5})$ for quarks are still allowed. These are numbers making detection in an experiment very challenging.

   This does not mean that the massless dark photon cannot be searched for in the laboratory. We must look either to processes where SM particles heavier than the electron or the muon and the u or d quarks are involved—and the most severe astrophysical bounds do not apply—or physics where the dipole operator in Eq. (1.15) is between fermions of different flavors or very high-energy processes where the large scale Λ is partially compensated by the scaling of the dipole and radius operators in Eq. (1.14) and Eq. (1.16) and the overall contribution is less suppressed.

   For example, for a first generation quark taken to be a parton in a hadron collider, the limit at an energy scale of 10 TeV, is of $d_M^{u,d} \simeq 10^{-3}$ (see Fig. 2.3) which would give a deviation in the cross section within the reach of future machines. Similarly, for the electron, the limits in Fig. 2.2 show that a $d_M^e \simeq 10^{-6}$ is still allowed at the scale of 1 TeV and therefore accessible at future lepton colliders for the projected sensitivity. As much suppressed as these cross sections are, they are comparable with those of the case of the massive dark photon after the corresponding limits are taken into account (see Sect. 3.3.1).

   These, and others possibilities, are discussed in Sect. 2.4 where some of the proposed experiments to search for the massless dark photon are reviewed.

2.2 Limits on Milli-Charged Particles

Milli-charged particles arise, as discussed in the Sect. 1.1 of the Introduction, in the case of a massless dark photon because the rotation of the mixing term in Eq. (1.1) leaves the photon coupled to the dark sector particles χ with strength $\varepsilon e'$. Searches are accordingly parameterized in terms of the mass m_χ and the electromagnetic coupling (modulated by ε) of the supposedly milli-charged dark-sector particle.

The physics of stellar evolution for horizontal branches, red giants, and white dwarves (RGWD [32]), together with supernovae (SN1987 [33]) provide bounds in the region of small masses ($m_\chi \lesssim 1$ MeV). In this region constraints on N_{eff} during nucleosynthesis and in the cosmic microwave background (N_{eff} BBN and CMB [32]) limits the possibility of having milli-charged particles. These limits are derived along the same lines discussed in the case of the massless dark photon.

Further limits can be derived from precision measurements in QED, notably from the Lamb shift in the transition $2S_{1/2}$-$2P_{3/2}$ in the Hydrogen atom [45] and the non-observation of the invisible decay of ortho-positronium (oPS [34]). Limits in the intermediate mass range $1 - 100$ MeV come from a SLAC dedicated experiment (SLAC milliQ [35]) and from the reinterpretation of data from the neutrino experiments LSND and miniBooNE [36].

Searches at LEP [37] and LHC [38] cover larger values of the mass (100 MeV $\lesssim m_\chi \lesssim 1$ TeV).

Finally, for very large masses ($m_\chi \gtrsim 10$ TeV) the impact on the cosmological parameters severely restricts the possible values of milli-charges (WMAP and dark matter relic density constraint, [38] and references therein).

All these limits are shown as filled area in the plot of Fig. 2.4.

Milli-charged particles as dark matter have been proposed (see for example [46] and [47]) to explain the anomalous 21 cm hydrogen absorption signal reported by the EDGES experiment [48]. Given the preliminary nature of the results, we have not included them in Fig. 2.4.

The projected limits of future experiments are depicted in Fig. 2.5 together with the current limits in gray background to show the expected advances. Of these, the most significant for masses around 1 GeV comes from the proposed milliQAN experiment [42] proposed to be installed on the surface above one of the LHC interaction points. MilliQAN could improve the collider limits by two orders of magnitude. The range in mass between 10-100 MeV can be optimally covered by the FerMINI experiment [41] proposed in the DUNE near detector hall at Fermilab. Finally the search for milli-charged particles below 10 MeV mass may be improved by almost two orders of magnitude by the LDMX experiment [43] proposed both at CERN [49] and at SLAC [50].

2.3 A Minimal Model of the Dark Sector

As discussed in Chap. 1, it is useful to underpin the phenomenology of the massless dark photon to a UV model. We consider a *minimal* model consisting of dark fermions that are, by definition, singlets under the SM gauge interactions. These dark fermions interact with the visible sector through a portal provided by scalar messengers which carry both SM and dark-sector charges. These scalars are phenomenologically akin to the sfermions of supersymmetric models.

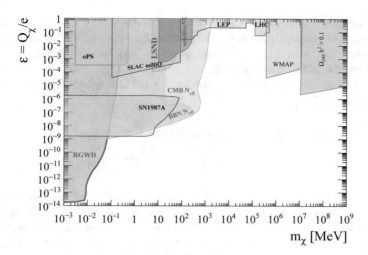

Fig. 2.4 Existing limits for experiments on milli-charged dark-sector matter. Limits from stellar evolution (RGWD [32] and SN1987A [33]); N_{eff} during nucleosynthesis and in the cosmic microwave background [32]; invisible decays of ortho-positronium (oPS) [34]; SLAC milliQ experiment [35]; reinterpretation of data from LSND and miniBooNE [36]; searches at LEP [37] and LHC [38]; WMAP results and dark matter relic density abundance [38]

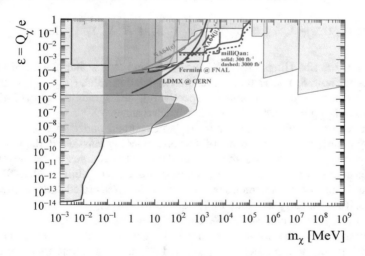

Fig. 2.5 Future sensitivities for proposed experiments on milli-charged dark-sector matter. Future sensitivities of NA64(e)$^{++}$ [39]; NA64(μ) [40]; FerMINI [41]; milliQAN [42]; LDMX [43]. The sensitivity shown for LDMX @ CERN corresponds to 10^{16} electrons-on-target and a beam energy of 16 GeV. The existing limits are shown as gray areas. The plot is revised from [44]

In general, we can have as many dark fermions as there are in the SM; they can be classified conveniently according to whether they couple (via the corresponding messengers) to quarks (q_L, u_R, d_R) or leptons (l_L, e_R): We denote the first (hadron-like) Q and the latter (lepton-like) χ.

The Yukawa-like interaction Lagrangian for flavor-diagonal interactions can be written as [51, 52]:

$$\mathcal{L} \supset -g_L \left(\phi_L^\dagger \bar{\chi}_R l_L + S_L^{U\dagger} \bar{Q}_R^U q_L + S_L^{D\dagger} \bar{Q}_R^D q_L \right)$$
$$- g_R \left(\phi_R^\dagger \bar{\chi}_L e_R + S_R^{U\dagger} \bar{Q}_L^D u_R + S_R^{U\dagger} \bar{Q}_L^D d_R \right) + \text{h.c.} \qquad (2.46)$$

where q_L (q_R) and e_L (e_R) are $SU(2)_L$ doublets (singlets) for quarks and leptons respectively. Sum over flavor and color indices, that we omitted for simplicity, is understood. The L-type scalars are doublets under $SU(2)_L$, while the R-type scalars are singlets under $SU(2)_L$. The $S_{L,R}$ messengers carry color indices (unmarked in (2.46)), while the messengers $\phi_{L,R}$ are color singlets. The Yukawa coupling strengths are parameterized by $\alpha_{L,R} \equiv g_{L,R}^2/(4\pi)$; they can be different for different fermions and as many as the SM fermions. For simplicity, we take them to be equal and, in addition, $\alpha_L = \alpha_R$.

In order to generate chirality-changing processes, we must have the mixing terms

$$\mathcal{L} \supset -\lambda_s S_0 \left(H^\dagger \phi_R^\dagger \phi_L + \tilde{H}^\dagger S_R^{U\dagger} S_L^U + H^\dagger S_R^{D\dagger} S_L^D \right) + \text{h.c.}, \qquad (2.47)$$

where H is the SM Higgs boson, $\tilde{H} = i\sigma_2 H^\star$, and S_0 a scalar singlet of the dark sector. After both S_0 and H take a vacuum expectation value (μ_S and v_h—the electroweak vacuum expectation value—respectively), the Lagrangian in Eq. (2.47) gives rise to the mixing between right- and left-handed states.

Dark sector and messenger states are both charged under an unbroken $U(1)_D$ gauge symmetry which is the same of the corresponding massless dark photon, with coupling strength α_D. We assign different dark $U(1)_D$ charges to the various dark sector fermions to ensures, by charge conservation, their stability. Since SM fields are neutral under $U(1)_D$ interactions, messengers and associated dark-fermions field in Eq. (2.46) must carry the same $U(1)_D$ quantum charge.

When the dark sector scalar S_0 and the Higgs boson acquire their vacuum expectation values, the scalar messengers must be rotated to identify the physical states. Before this rotation, $\phi_{L\nu}$, S_{Ld}^U, and S_{Lu}^U are degenerate mass eigenstates with mass m_S. After the rotation, the mass eigenstates (labeled by $\pm$) are given by

$$\phi_\pm \equiv \frac{1}{\sqrt{2}} \left(\phi_L \pm \phi_R \right), \quad S_\pm^{U,D} \equiv \frac{1}{\sqrt{2}} \left(S_L^{U,D} \pm S_R^{U,D} \right), \text{ corresponding to masses}$$

$$m_\pm = m_{\phi,S} \sqrt{1 \pm \eta_s} \qquad (2.48)$$

where we defined the mixing parameters for the S and ϕ messengers

$$\eta_{\phi,S} \equiv \frac{\lambda_s \mu_{\phi,S} v_h}{m_S^2}.$$ (2.49)

In the new basis, the interaction terms in Eq. (2.46) in the lepton sector is given by

$$\mathcal{L}^{(lep)} \supset -g_L \phi_{L\nu}^{\dagger} (\bar{\chi}_R \nu_L) - \frac{g_L}{\sqrt{2}} \left(\phi_+^{\dagger} + \phi_-^{\dagger} \right) (\bar{\chi}_R e_L)$$

$$- \frac{g_R}{\sqrt{2}} \left(\phi_+^{\dagger} - \phi_-^{\dagger} \right) (\bar{\chi}_L e_R) + \text{h.c.}.$$ (2.50)

The corresponding interaction terms in the hadronic sector have the same form.

Looking at (2.50), we can see that if χ is a stable dark-sector species, then its mass must be at most $m_- + m_e$. Similarly, for a dark-sector species Q, the mass must be no heavier than $m_- + m_q$, where m_q is the mass of the SM species corresponding to Q. This sets an upper bound for the mixing $\eta_{\phi,S}$:

$$\eta_{\phi,S} < 1 - \left(\frac{M}{m_{\phi,S}} \right)^2.$$ (2.51)

In Eq. (2.51), M is the mass of the heaviest stable dark-sector species. We assume that M is heavier than any SM species. The upper bound in Eq. (2.51) also guarantees that the scalar messengers are heavier than the dark fermion into which they can thus decay.

This model can be considered as a template for many models of the dark sector with the scalar messenger as stand-in for more complicated portals. It is a simplified version of the model in [51], which might provide a natural solution to the SM flavor-hierarchy problem.

The discussion above is restricted to the flavor-diagonal interactions. A more general flavor structure in the portal interaction, including the off-diagonal terms, arising as a consequence of the simultaneous diagonalization of the dark-fermion mass and quark interaction basis, can be simply obtained by generalizing the above terms as follows [53]

$$S_L^{U_i\dagger} \bar{Q}_R^{U_i} q_L^i \rightarrow S_L^{U_i\dagger} \bar{Q}_R^{U_i} (\rho_L^U)_{ij} q_L^j$$
$$S_R^{U_i\dagger} \bar{Q}_L^{U_i} q_R^i \rightarrow S_R^{U_i\dagger} \bar{Q}_L^{U_i} (\rho_L^U)_{ij} q_R^j,$$ (2.52)

and analogously for the down and lepton sectors, where i, j are explicit flavor indices and sum over i, j is understood.

To keep the contribution to the dipole coefficient simple, lest the generality obfuscates the estimate, we follow the guidelines of the model in [51]. We assume that the masses of the messengers ϕ^i, $S_{L,R}^{U,i}$ and $S_{L,R}^{D,i}$ are the same and the mixing matrices ρ_{ij} have a hierarchical structure (like in the SM) with the off-diagonal smaller than the diagonal terms. The former hypothesis is a consequence of the $SU(N_F)$ flavor

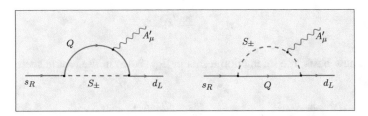

Fig. 2.6 Vertex diagrams for the generation of the dipole operators in the model of the dark sector

symmetry in the free lagrangian of messenger sector (with $N_F = 6$)) [51], while the latter follows from the requirement of minimal flavor violation hypothesis [52].

We also take $\rho_{ij} \equiv \rho^D_{ij} = \rho^U_{ij}$. This way, the loop of dark sector particles is dominated by the contribution with the heaviest dark fermion coupled to the SM fermions of flavor i and j with one coefficient off-diagonal ρ_{ij} and one diagonal ρ_{ii}. In the following, in order to distinguish the contribution from the up and down sector couplings we will use the notation $\rho_{uu} \equiv \rho^U_{11}$, $\rho_{dd} \equiv \rho^D_{11}$, $\rho_{sd} \equiv \rho^D_{21}$, and similarly for the other coefficients.

Matching the model (Fig. 2.6) to the effective Lagrangian given in Eq. (1.14) after integrating the loop, and identifying the scale Λ as

$$\frac{v_h}{\Lambda^2} \simeq \frac{m_{Q^i}}{m_S^2}, \tag{2.53}$$

with m_{Q^i} the heaviest dark-fermion running in the loop, we can re-express the magnetic dipole explicitly in terms of the parameters of the model. For example, in the case of the generic (quark) flavor transition from $i \rightarrow j$, with i- and j mixing, neglecting the SM masses, according to the Lagrangian in (2.46) and substitutions (2.52), we have [53]

$$\mathbb{D}^{ij}_M = \rho_{jj}\rho^*_{ij} \, \Re \left[\frac{g_L g_R}{(4\pi)^2} \right] F_M(x, \eta_s) . \tag{2.54}$$

where $x = (m_{Q^i})^2/m_S^2$ and η_s the mixing parameter defined in (2.49). In the following, we will introduce the notation of m_{S^U} and m_{S^D} to distinguish the common messenger mass in the up and down $SU(2)_L$ sectors respectively, and $\eta_s^{U,D}$ for the corresponding mixing parameters. The function $F_M(x, y)$ is given by [53]

$$F_M(x, y) = \frac{1}{2}\Big[f(x, y) - f(x, -y) \Big], \tag{2.55}$$

where

$$f(x, y) = \frac{1 - x + y + (1 + y)\log\left(\frac{x}{1+y}\right)}{(1 - x + y)^2}. \tag{2.56}$$

CP-violating phases, relevant for flavor changing processes, can arise from the mixing parameters. For instance, in the $n \to m$ flavor transition, we can have CP-violating phase δ_{CP} from the relation

$$\rho_{nm}\rho_{mm}^* - \rho_{nm}^*\rho_{mm} = 2i\sin\delta_{CP}. \tag{2.57}$$

2.3.1 Constraints on the UV Model Parameters

The introduction of the UV model makes possible to re-discuss the bounds of Sect. 2.1 on the massless dark photon in terms of the parameters of the model.

There are no laboratory limits for the masses of the dark fermions from events in which they are produced because they are SM singlets and do not interact directly with the detector. Cosmological bounds have been considered in [54] where, in particular, avoiding distortions of the cosmic microwave background is shown to require the masses of the dark fermions to be larger than 1 GeV or, if lighter, that the coupling α_L and α_R be less than 10^{-3}.

The messenger states have the same quantum numbers and spin as the supersymmetric squarks. At the LHC they are copiously produced in pairs through QCD interactions and decay at tree level into a quark and a dark fermion. The final state arising from their decay is thus the same as the one obtained from the $\tilde{q} \to q\chi_1^0$ process. Therefore limits on the messenger masses can be obtained by reinterpreting supersymmetric searches on first and second generation squarks decaying into a light jet and a massless neutralino [55], assuming that the gluino is decoupled. A lower bound on their masses is thus obtained [56] to give

$$m_S^i \gtrsim 940 \text{ GeV}, \tag{2.58}$$

for the messenger mass related to the dark fermions Q^U and Q^D. This limit increases up to 1.5 TeV by assuming that messengers of both chiralities associated to the first and second generation of SM quarks are degenerate in mass.

For the masses of the lepton-like scalar messengers, constraints on the mass of sleptons [57] give the following lower bound on the messenger mass in the lepton sector:

$$m_\phi \gtrsim 290 \text{ GeV}. \tag{2.59}$$

All the limits discussed in Sect. 2.1 can be re-expressed in terms of the UV model parameters.

For example, the limit from stellar cooling in Eq. (2.9) becomes

$$\frac{m_\phi^2/m_{\chi^e}}{\sqrt{\alpha_D\alpha_L\alpha_R}\,|\rho_{ee}|^2|\,F_M(x_e,\eta_\phi)} \gtrsim 2.1 \times 10^6 \text{ TeV}, \tag{2.60}$$

where $x_e = m_\phi^2/m_{\chi^e}^2$, with m_{χ^e} the dark fermion mass associated to the electron, and η_ϕ the corresponding mixing parameter in the colorless messengers sector, and the loop function $F_M(x, y)$ is given in Eq. (2.55). This limit, which is obtained by rescaling the right-hand side of Eq. (2.9) for $1/(4\pi v_h)$, applies specifically to the Yukawa coupling of electrons and the corresponding messenger state.

For a quick estimate of the bound above and those that follow, the loop function $F_M(x, y)$ can be considered a coefficient of order $O(10^{-1})$ as long as η_ϕ is not too small. For instance, for $x \simeq 1$ and $y \simeq 0.5$, the loop function $F_M \simeq 0.09$.

Similarly, by using the same rescaling factor, the neutrino signal of supernova 1987A and the limit in Eq. (2.15) yields now

$$\frac{m_S^2/m_{Q^u}}{\sqrt{\alpha_D \alpha_L \alpha_R}\, |\rho_{uu}|^2\, F_M(x_u, \eta_S)} \gtrsim 2.0 \times 10^5 \text{ TeV}, \tag{2.61}$$

where now $x_u = m_S^2/m_{Q^u}^2$, with m_{Q^u} the dark-fermion associated to the light u quark. A similar limit holds for the case of the d quark sector.

The others bounds in Sect. 2.1 can be written in terms of the parameters of the model in the same way.

Instead, new bounds can be set now that we have un underlying UV model because the scalar messengers carry also the electromagnetic charge. Processes with the visible photon can thus be used; these processes were not available for the model-independent case in Sect. 2.1 for which only the coupling to the dark photon was taken into account.

The magnetic moment of the SM fermions arises from the one-loop diagram of the states of the UV model.

From Eq. (2.33) in Sect. 2.1, we find

$$\frac{m_\phi^2/m_{\chi^e}}{\sqrt{\alpha_L \alpha_R}\, |\rho_{ee}|^2\, G_M(x_e, \eta_\phi)} \gtrsim 9.8 \times 10^4 \text{ TeV}, \tag{2.62}$$

where $x_e = m_{\chi^e}^2/m_\phi^2$, with m_{χ^e} the dark-fermion mass associated to the muon. The loop function is in this case given by [53]

$$G_M(x, y) = \frac{1}{2}\Big[g(x, y) - g(x, -y)\Big], \tag{2.63}$$

where

$$g(x, y) = \frac{(1 + y)^2 - x^2 + 2x\,(1 + y)\log\left(\frac{x}{1+y}\right)}{2\,(x - 1 - y)^3}. \tag{2.64}$$

Also interesting is the anomalous magnetic moment of the muon because of the lingering discrepancy between theory and experiments. From Eq. (2.35) in Sect. 2.1, we find

$$\frac{m_\phi^2/m_{\chi^\mu}}{\sqrt{\alpha_L \alpha_R}\, |\rho_{\mu\mu}|^2\, G_M(x_\mu, \eta_\phi)} \gtrsim 6.3 \times 10^3 \text{ TeV}, \tag{2.65}$$

where $x_\mu = m_{\chi^\mu}^2/m_\phi^2$, with m_{χ^μ} the dark-fermion mass associated to the muon. Again, for a quick estimate of the bounds above and those that follow, the loop function $G_M(x, y)$ can be considered a coefficient of order $O(10^{-1})$ as long as η_ϕ is not too small. For instance, for $x \simeq 1$ and $y \simeq 0.5$, the loop function $G_M \simeq 0.05$.

The various Yukawa couplings and messenger and fermion masses are probed in a selective manner in flavor physics where we must distinguish among the various couplings and states. Mixing (proportional to a coefficient ρ_{ij} in the equations below) between different flavor states must be included.

The strongest bound comes from the limit on the BR $(\mu \to e\gamma) < 4.2 \times 10^{-13}$ (CL 90%) [58] of the MEG experiment. From this result, we find that

$$\frac{m_\phi^2/m_{\chi^\mu}}{\sqrt{\alpha_L \alpha_R}\, |\rho_{\mu\mu}\rho_{\mu e}^\star|\, G_M(x_\mu, \eta_\phi)} \gtrsim 4.9 \times 10^8 \text{ TeV}. \tag{2.66}$$

A weaker bound can be extracted, in the hadronic sector, from the difference between the experimental limit on the BR $(B \to X_s\gamma) < (3.21 \pm 0.33) \times 10^{-4}$ [59] of the BaBar collaboration and its SM estimate [60]. It yields

$$\frac{m_S^2/m_{Q^b}}{\sqrt{\alpha_L \alpha_R}\, |\rho_{bb}\rho_{bs}^\star|\, G_M(x_b, \eta_S)} \gtrsim 1.3 \times 10^4 \text{ TeV}, \tag{2.67}$$

where $x_b = m_{Q^b}^2/m_S^2$, with m_{Q^b} the mass of dark fermion associated to the b-quark,

The limits in Eqs. (2.66) and (2.67) apply specifically to the off-diagonal terms in the Yukawa couplings ρ_{ij} of the muon-electron and b-s quark mixing respectively, and to the corresponding mass of messenger states.

The mass mixing in the Kaon system [53, 61] gives a further limit

$$\frac{m_S^2}{(\alpha_L^2 + \alpha_R^2)\, |\rho_{ss}\rho_{sd}^\star|^2} \gtrsim 3 \times 10^5 \text{ TeV}^2, \tag{2.68}$$

which is not related to the dark photon and its coupling α_D because it comes from the box-diagram insertion of the dark scalars and fermions.

The limit in Eq. (2.68) is obtained by requiring that the messenger contribution to the box diagram for the K^0-$\bar{K}^0$ mixing does not exceed the experimental value of the mixing parameter $\Delta m_K = 3.48 \times 10^{-12}$MeV [62]. Due to chirality arguments, the leading contribution to the box diagram in Eq. (2.68) does not depend on the dark fermion mass, which is assumed to be much smaller than the corresponding messenger mass in the down sector and therefore very weakly on the loop function.

The limit in Eq. (2.68) applies specifically to the off-diagonal term in the Yukawa coupling of d-s quark mixing and the corresponding messenger state. A similar but weaker bound can be found from B-meson mixing.

As displayed in the equations above, all these limit can be made weaker by taking m_χ (or m_Q) sufficiently light or by varying the corresponding mixing parameters η_s, η_ϕ. In the UV model is thus possible to play with the parameters to make room for larger values of the dipole coefficient by absorbing part of the suppression in the connection between the scale Λ and the mass ratios m_χ/m_ϕ^2 and m_Q/m_S. For instance a scale $\Lambda = 1$ TeV for the new physics of the dark sector is still allowed by the stringent bound in Eq. (2.35) if we take m_χ sufficiently small. This way, there is some additional freedom in comparing limits from different processes as compared to the model-independent case where the scale Λ is taken to be the same for all bounds.

2.4 Future Experiments

The massless dark photon has been neglected so far from the experimental point of view as compared to the massive one. It is one of the aims of the present review to boost the community scrutiny in this direction. In the past few year several proposals have been put forward and new experiments are in the planning:

- Flavor physics: This is one of the most promising areas for searching for the dark photon and the dark sector in general because none of the stringent astrophysical constrains discussed in Sect. 2.1 applies given the flavor off-diagonal nature of the dipole operator in these cases.
  Proposals exist for processes in Kaon physics at NA62 [63]. The Kaon decay $K \to \pi A'$ is forbidden by the conservation of angular momentum but the decay $K^+ \to \pi^0 \pi^+ A'$ is allowed and the estimated branching ratio [61] is within reach of the current sensitivity. The rare decays $K^+ \to \pi^+ \nu \bar{\nu}$ [64] and $K_L \to \pi^0 \nu \bar{\nu}$ [65] are other two processes where the physics of the dark photon can play a crucial role [66]. Also Hyperion decays can be used for detecting the production of A' [67] and in the decay of charmed hadrons [68] and BESIII.
  In addition, decays into invisible states of B-mesons at BaBar [69] and Belle [70] and $K_{L,S}$ and other neutral mesons at NA64 [71, 72] can be used to study the dark sector (assuming the invisible states belong to it). These decays are greatly enhanced by the Fermi-Sommerfeld [73, 74] effect due to their interaction with the dark photon—the same way as ordinary decays, like the β-decay, are enhanced by the same effect—making this another exciting area for searching the dark sector [56].
- Higgs and Z physics: The striking signature of a mono-photon plus missing energy can be used to search Higgs [75–77] and Z-boson [78, 79] decay into a visible and a dark photon. Again, the stringent astrophysical constrains discussed in Sect. 2.1 do not apply because the size of the dipole operator is dominated (in the loop diagram) by the heavy-quark contribution's giving raise to the coupling to the dark photon, as discussed in Sect. 2.3.

- Pair annihilation: Collider experiment at higher energies and luminosities can use the same striking signature of a mono-photon plus missing energy to search for the dark photon. Even though the dipole interaction is suppressed and severely constrained in this case by the astrophysical and cosmological bounds discussed in Sect. 2.1, it is no more suppressed than the equivalent cross sections for the massive case. Moreover, the dipole operator scales as the center-of-mass energy in the process and higher energies make it more and more relevant;
- Magnons: An interesting possibility is the use of magnons in ferromagnetic materials and their interaction with dark photons (QUAX proposal) [80, 81]. The estimated sensitivity is again done for axions but can be translated for massless dark photons as in the discussion about stars above.
- Astrophysics: Gravitation waves emitted during the inspiral phase of neutron star collapse can test the presence of other forces beside gravitation. Dipole radiation by even small amount of charges on the stars modifies the energy emitted; the dark photon is a prime candidate for this kind of correction [82–85].

References

1. S. Hoffmann, Paraphotons and Axions: similarities in stellar emission and detection. Phys. Lett. B **193**, 117–122 (1987). https://doi.org/10.1016/0370-2693(87)90467-9
2. G.G. Raffelt, Stars as laboratories for fundamental physics, vol. 664 (University Press, Chicago, USA, 1996), p. 1996. http://wwwth.mpp.mpg.de/members/raffelt/mypapers/199613.pdf
3. E.D. Carlson, Limits on a new U(1) coupling. Nucl. Phys. B **286**, 378–398 (1987). https://doi.org/10.1016/0550-3213(87)90446-9
4. B.A. Dobrescu, Massless gauge bosons other than the photon. Phys. Rev. Lett. **94**, 151802 (2005). https://doi.org/10.1103/PhysRevLett.94.151802. arXiv:hep-ph/0411004 [hep-ph]
5. M. Nakagawa, Y. Kohyama, N. Itoh, Axion bremsstrahlung in dense stars. Astrophys. J. **322**, 291 (1987). https://doi.org/10.1086/165724
6. G.G. Raffelt, Axion bremsstrahlung in red giants. Phys. Rev. D **41**, 1324–1326 (1990). https://doi.org/10.1103/PhysRevD.41.1324
7. M.M. Miller Bertolami, B.E. Melendez, L.G. Althaus, J. Isern, Revisiting the Axion bounds from the Galactic white dwarf luminosity function. JCAP **1410** (10), 069 (2014). arXiv:1406.7712 [hep-ph]. https://doi.org/10.1088/1475-7516/2014/10/069
8. N. Viaux, M. Catelan, P.B. Stetson, G. Raffelt, J. Redondo, A.A.R. Valcarce, A. Weiss, Neutrino and Axion bounds from the globular cluster M5 (NGC 5904). Phys. Rev. Lett. **111**, 231301 (2013). https://doi.org/10.1103/PhysRevLett.111.231301. arXiv:1311.1669 [astro-ph.SR]
9. M. Giannotti, I. Irastorza, J. Redondo, A. Ringwald, Cool WISPs for stellar cooling excesses. JCAP **1605**(05), 057 (2016). arXiv:1512.08108 [astro-ph.HE]. https://doi.org/10.1088/1475-7516/2016/05/057
10. R.P. Brinkmann, M.S. Turner, Numerical rates for Nucleon-Nucleon Axion Bremsstrahlung. Phys. Rev. D **38**, 2338 (1988). https://doi.org/10.1103/PhysRevD.38.2338
11. G. Raffelt, D. Seckel, A selfconsistent approach to neutral current processes in supernova cores. Phys. Rev. D **52**, 1780–1799 (1995). arXiv:astro-ph/9312019. https://doi.org/10.1103/PhysRevD.52.1780
12. N. Iwamoto, Axion emission from neutron stars. Phys. Rev. Lett. **53**, 1198–1201 (1984). https://doi.org/10.1103/PhysRevLett.53.1198

13. W. Keil, H.-T. Janka, D.N. Schramm, G. Sigl, M.S. Turner, J.R. Ellis, A fresh look at Axions and SN-1987A. Phys. Rev. D **56**, 2419–2432 (1997). https://doi.org/10.1103/PhysRevD.56. 2419.arXiv:astro-ph/9612222 [astro-ph]
14. P. Carenza, T. Fischer, M. Giannotti, G. Guo, G. Martínez-Pinedo, A. Mirizzi, Improved Axion emissivity from a supernova via nucleon-nucleon bremsstrahlung. JCAP **1910**(10), 016 (2019). arXiv:1906.11844 [hep-ph]. https://doi.org/10.1088/1475-7516/2019/10/016
15. B.D. Fields, K.A. Olive, T.-H. Yeh, C. Young, Big-Bang nucleosynthesis after planck. JCAP **2003**(03), 010 (2020). arXiv:1912.01132 [astro-ph.CO]. https://doi.org/10.1088/1475-7516/2020/03/010
16. B.A. Dobrescu I. Mocioiu, Spin-dependent macroscopic forces from new particle exchange. JHEP **11**, 005 (2006). arXiv:hep-ph/0605342 [hep-ph]
17. F. Ficek, D.F.J. Kimball, M. Kozlov, N. Leefer, S. Pustelny, D. Budker, Constraints on exotic spin-dependent interactions between electrons from helium fine-structure spectroscopy. Phys. Rev. **A95**(3), 032505 (2017). arXiv:1608.05779 [physics.atom-ph]. https://doi.org/10.1103/PhysRevA.95.032505
18. W.-T. Ni, S.-S. Pan, H.-C. Yeh, L.-S. Hou, J.-L. Wan, Search for an Axionlike spin coupling using a paramagnetic salt with a dc SQUID. Phys. Rev. Lett. **82**, 2439–2442 (1999). https://doi.org/10.1103/PhysRevLett.82.2439
19. D.J. Wineland, J.J. Bollinger, D.J. Heinzen, W.M. Itano, M.G. Raizen, Search for anomalous spin-dependent forces using stored-ion spectroscopy. Phys. Rev. Lett. **67**, 1735–1738 (1991). https://doi.org/10.1103/PhysRevLett.67.1735
20. D. Hanneke, S. Fogwell, G. Gabrielse, New measurement of the electron magnetic moment and the fine structure constant. Phys. Rev. Lett. **100**, 120801 (2008). https://doi.org/10.1103/PhysRevLett.100.120801. arXiv:0801.1134 [physics.atom-ph]
21. G. Giudice, P. Paradisi, M. Passera, Testing new physics with the electron g-2. JHEP **11**, 113 (2012). https://doi.org/10.1007/JHEP11(2012)113. arXiv:1208.6583 [hep-ph]
22. Muon g-2 Collaboration, G. W. Bennett et al., Final report of the muon E821 anomalous magnetic moment measurement at BNL. Phys. Rev. **D73**, 072003 (2006). arXiv:hep-ex/0602035 [hep-ex]. https://doi.org/10.1103/PhysRevD.73.072003
23. RBC, UKQCD Collaboration, T. Blum, P. Boyle, V. Gülpers, T. Izubuchi, L. Jin, C. Jung, A. Jüttner, C. Lehner, A. Portelli, J. Tsang, Calculation of the hadronic vacuum polarization contribution to the muon anomalous magnetic moment. Phys. Rev. Lett. **121**(2), 022003 (2018). arXiv:1801.07224 [hep-lat]. https://doi.org/10.1103/PhysRevLett.121.022003
24. TWIST Collaboration, R. Bayes et al., Search for two body muon decay signals. Phys. Rev. D **91**(5), 052020 (2015). https://doi.org/10.1103/PhysRevD.91.052020 arXiv:1409.0638 [hep-ex]
25. NA62 Collaboration, J. Engelfried, Search for $K^+ \to \pi^+ \nu\bar{\nu}$: first NA62 results. Springer Proc. Phys. **234**, 135–141 (2019). DOI: https://doi.org/10./978-3-030-29622-3_19
26. J. Engel, D. Seckel, A.C. Hayes, Emission and detectability of hadronic Axions from SN1987A. Phys. Rev. Lett. **65**, 960–963 (1990). https://doi.org/10.1103/PhysRevLett.65.960
27. OPAL Collaboration, G. Abbiendi et al., Search for anomalous photonic events with missing energy in e^+e^- collisions at $\sqrt{s} = 130$-GeV, 136-GeV and 183-GeV. Eur. Phys. J. **C8**, 23–40 (1999). arXiv:hep-ex/9810021 [hep-ex]. https://doi.org/10.1007/s100520050442
28. L3 Collaboration, M. Acciarri et al., Single and multiphoton events with missing energy in e^+e^- collisions at $\sqrt{s}$–189-GeV. Phys. Lett. **B470**, 268–280 (1999). arXiv:hep-ex/9910009 [hep-ex]. https://doi.org/10.1016/S0370-2693(99)01286-1
29. ALEPH Collaboration, A. Heister et al., Single photon and multiphoton production in e^+e^- collisions at $\sqrt{s}$ up to 209-GeV. Eur. Phys. J. **C28**, 1–13 (2003). https://doi.org/10.1140/epjc/s2002-01129-7
30. ATLAS Collaboration, M. Aaboud et al., Search for new phenomena in events with a photon and missing transverse momentum in pp collisions at $\sqrt{s} = 13$ TeV with the ATLAS detector. JHEP **06**, 059 (2016). arXiv:1604.01306 [hep-ex]. https://doi.org/10.1007/JHEP06(2016)059
31. C.M.S. Collaboration, A.M. Sirunyan et al., Search for new physics in final states with a single photon and missing transverse momentum in proton-proton collisions at $\sqrt{s} = 13$ TeV. JHEP **02**, 074 (2019). https://doi.org/10.1007/JHEP02(2019)074. arXiv:1810.00196 [hep-ex]

32. H. Vogel, J. Redondo, Dark Radiation constraints on minicharged particles in models with a hidden photon. JCAP **1402**, 029 (2014). https://doi.org/10.1088/1475-7516/2014/02/029. arXiv:1311.2600 [hep-ph]

33. J.H. Chang, R. Essig, S.D. McDermott, Supernova 1987A constraints on Sub-GeV dark sectors, millicharged particles, the QCD Axion, and an Axion-like particle. JHEP **09**, 051 (2018). https://doi.org/10.1007/JHEP09(2018)051. arXiv:1803.00993 [hep-ph]

34. A. Badertscher, P. Crivelli, W. Fetscher, U. Gendotti, S. Gninenko, V. Postoev, A. Rubbia, V. Samoylenko, D. Sillou, An improved limit on invisible decays of positronium. Phys. Rev. D **75**, 032004 (2007). https://doi.org/10.1103/PhysRevD.75.032004. arXiv:hep-ex/0609059 [hep-ex]

35. A.A. Prinz et al., Search for millicharged particles at SLAC. Phys. Rev. Lett. **81**, 1175–1178 (1998). arXiv:hep-ex/9804008 [hep-ex]. https://doi.org/10.1103/PhysRevLett.81.1175

36. G. Magill, R. Plestid, M. Pospelov, Y.-D. Tsai, Millicharged particles in neutrino experiments. Phys. Rev. Lett. **122**(7), 071801 (2019). arXiv:1806.03310 [hep-ph]. https://doi.org/10.1103/PhysRevLett.122.071801

37. S. Davidson, S. Hannestad, G. Raffelt, Updated bounds on millicharged particles. JHEP **05**, 003 (2000). https://doi.org/10.1088/1126-6708/2000/05/003. arXiv:hep-ph/0001179 [hep-ph]

38. J. Jaeckel, M. Jankowiak, M. Spannowsky, LHC probes the hidden sector. Phys. Dark Univ. **2**, 111–117 (2013). https://doi.org/10.1016/j.dark.2013.06.001. arXiv:1212.3620 [hep-ph]

39. D. Banerjee et al., Addendum to the NA64 Proposal: search for $A' \to$ Invisible and $X \to e^+e^-$ Decays in 2021. http://cds.cern.ch/record/2300189?ln=en

40. D. Banerjee et al., Addendum to the Proposal P348: Search for Dark Sector Particles Weakly Coupled to Muon with NA64 μ. http://cds.cern.ch/record/2640930?ln=en

41. K.J. Kelly, Y.-D. Tsai, Proton fixed-target scintillation experiment to search for millicharged dark matter. Phys. Rev. **D100**(1), 015043 (2019). arXiv:1812.03998 [hep-ph]. https://doi.org/10.1103/PhysRevD.100.015043

42. A. Ball et al., A Letter of Intent to Install a Milli-Charged Particle Detector at LHC P5. arXiv:1607.04669 [physics.ins-det]

43. LDMX Collaboration, T. Akesson et al., Light Dark Matter eXperiment (LDMX). arXiv:1808.05219 [hep-ex]

44. J. Beacham et al., Physics beyond colliders at CERN: beyond the standard model working group report. J. Phys. G **47**(1), 010501 (2020). arXiv:1901.09966 [hep-ex]. https://doi.org/10.1088/1361-6471/ab4cd2

45. E.W. Hagley, F.M. Pipkin, Separated oscillatory field measurement of hydrogen S-21/2- P-23/2 fine structure interval. Phys. Rev. Lett. **72**, 1172–1175 (1994). https://doi.org/10.1103/PhysRevLett.72.1172

46. E.D. Kovetz, V. Poulin, V. Gluscevic, K.K. Boddy, R. Barkana, M. Kamionkowski, Tighter limits on dark matter explanations of the anomalous EDGES 21 cm signal. Phys. Rev. **D98**(10), 103529 (2018). arXiv:1807.11482 [astro-ph.CO]. https://doi.org/10.1103/PhysRevD.98.103529

47. H. Liu, N.J. Outmezguine, D. Redigolo, T. Volansky, Reviving Millicharged dark matter for 21-cm Cosmology. Phys. Rev. D. **100**(12), 123011 (2019). https://doi.org/10.1103/PhysRevD.100.123011. arXiv:1908.06986 [hep-ph]

48. R.A. Monsalve, B. Greig, J.D. Bowman, A. Mesinger, A.E.E. Rogers, T.J. Mozdzen, N.S. Kern, N. Mahesh, Results from EDGES high-band: II. Constraints on parameters of early galaxies. Astrophys. J. **863**(1), 11 (2018). https://doi.org/10.3847/1538-4357/aace54. arXiv:1806.07774 [astro-ph.CO]

49. T. Akesson et al., Dark Sector Physics with a Primary Electron Beam Facility at CERN

50. T. Raubenheimer, A. Beukers, A. Fry, C. Hast, T. Markiewicz, Y. Nosochkov, N. Phinney, P. Schuster, N. Toro, DASEL: Dark Sector Experiments at LCLS-II. arXiv:1801.07867 [physics.acc-ph]

51. E. Gabrielli, M. Raidal, Exponentially spread dynamical Yukawa couplings from nonperturbative chiral symmetry breaking in the dark sector. Phys. Rev. **D89**(1), 015008 (2014). arXiv:1310.1090 [hep-ph]. https://doi.org/10.1103/PhysRevD.89.015008

52. E. Gabrielli, L. Marzola, M. Raidal, Radiative Yukawa couplings in the simplest left-right symmetric model. Phys. Rev. **D95**(3), 035005 (2017). arXiv:1611.00009 [hep-ph]. https://doi.org/10.1103/PhysRevD.95.035005
53. E. Gabrielli, B. Mele, M. Raidal, E. Venturini, FCNC decays of standard model fermions into a dark photon. Phys. Rev. **D94**(11), 115013 (2016). arXiv:1607.05928 [hep-ph]. https://doi.org/10.1103/PhysRevD.94.115013
54. J.T. Acuña, M. Fabbrichesi, P. Ullio, Phenomenological Consequences of an Interacting Multicomponent Dark Sector. arXiv:2005.04146 [hep-ph]
55. ATLAS Collaboration, M. Aaboud et al., Search for squarks and gluinos in final states with jets and missing transverse momentum using 36 fb^{-1} of $\sqrt{s} = 13$ TeV pp collision data with the ATLAS detector. Phys. Rev. **D97**(11), 112001 (2018). arXiv:1712.02332 [hep-ex]. https://doi.org/10.1103/PhysRevD.97.112001
56. D. Barducci, M. Fabbrichesi, E. Gabrielli, Neutral Hadrons disappearing into the darkness. Phys. Rev. **D98** (3), 035049 (2018). arXiv:1806.05678 [hep-ph]. https://doi.org/10.1103/PhysRevD.98.035049
57. CMS Collaboration, A.M. Sirunyan et al., Search for supersymmetric partners of electrons and muons in proton-proton collisions at $\sqrt{s} = 13$ TeV. Phys. Lett. **B790**, 140–166 (2019). arXiv:1806.05264 [hep-ex]. https://doi.org/10.1016/j.physletb.2019.01.005
58. MEG Collaboration, A. Baldini et al., Search for the lepton flavour violating decay $\mu^+ \to e^+$ with the full dataset of the MEG experiment. Eur. Phys. J. C **76**(8), 434 (2016). arXiv:1605.05081 [hep-ex]. https://doi.org/10.1140/epjc/s10052-016-4271-x
59. BaBar Collaboration, J. Lees et al., Precision measurement of the $B \to X_s\gamma$ photon energy spectrum, branching fraction, and direct CP asymmetry $A_{CP}(B \to X_{s+d}\gamma)$. Phys. Rev. Lett. **109**, 191801 (2012). https://doi.org/10.1103/PhysRevLett.109.191801. arXiv:1207.2690 [hep-ex]
60. M. Misiak et al., Estimate of $\mathcal{B}(\bar{B} \to X_s\gamma)$ at $O(\alpha_s^2)$. Phys. Rev. Lett. **98**, 022002 (2007). https://doi.org/10.1103/PhysRevLett.98.022002. arXiv:hep-ph/0609232
61. M. Fabbrichesi, E. Gabrielli, B. Mele, Hunting down massless dark photons in kaon physics. Phys. Rev. Lett. **119**(3), 031801 (2017). arXiv:1705.03470 [hep-ph]. https://doi.org/10.1103/PhysRevLett.119.031801
62. Particle Data Group Collaboration, M. Tanabashi et al., Review of particle physics. Phys. Rev. D **98**(3), 030001 (2018). https://doi.org/10.1103/PhysRevD.98.030001
63. NA62 Collaboration, E. Cortina Gil et al., The Beam and detector of the NA62 experiment at CERN. JINST **12**(05), P05025 (2017). arXiv:1703.08501 [physics.ins-det]. https://doi.org/10.1088/1748-0221/12/05/P05025
64. NA62 Collaboration, E. Cortina Gil et al., First search for $K^+ \to \pi^+\nu\bar{\nu}$ using the decay-in-flight technique. Phys. Lett. B **791**, 156–166 (2019). arXiv:1811.08508 [hep-ex]. https://doi.org/10.1016/j.physletb.2019.01.067
65. KOTO Collaboration, J. Ahn et al., Search for the $K_L \to \pi^0\nu\bar{\nu}$ and $K_L \to \pi^0 X^0$ decays at the J-PARC KOTO experiment. Phys. Rev. Lett. **122**(2), 021802 (2019). arXiv:1810.09655 [hep-ex]. https://doi.org/10.1103/PhysRevLett.122.021802
66. M. Fabbrichesi, E. Gabrielli, Dark-sector physics in the search for the rare decays $K^+ \to \pi^+\nu\bar{\nu}$ and $K_L \to \pi^0\nu\bar{\nu}$. Eur. Phys. J. C **80**(6), 532 (2020). arXiv:1911.03755 [hep-ph]. https://doi.org/10.1140/epjc/s10052-020-8103-7
67. J.-Y. Su, J. Tandean, Searching for dark photons in hyperon decays. arXiv:1911.13301 [hep-ph]
68. J.-Y. Su, J. Tandean, Seeking massless dark photons in the decays of charmed hadrons. arXiv:2005.05297 [hep-ph]
69. BaBar Collaboration, J.P. Lees et al., Improved limits on B^0 decays to invisible final states and to $\nu\bar{\nu}\gamma$. Phys. Rev. **D86**, 051105 (2012). arXiv:1206.2543 [hep-ex]. https://doi.org/10.1103/PhysRevD.86.051105
70. Belle Collaboration, C.L. Hsu et al., Search for B^0 decays to invisible final states. Phys. Rev. **D86**, 032002 (2012). arXiv:1206.5948 [hep-ex]. https://doi.org/10.1103/PhysRevD.86.032002
71. S.N. Gninenko, N.V. Krasnikov, Invisible K_L decays as a probe of new physics. Phys. Rev. **D92**(3), 034009 (2015). arXiv:1503.01595 [hep-ph]. https://doi.org/10.1103/PhysRevD.92.034009

72. S.N. Gninenko, Search for invisible decays of π^0, η, η', K_S and K_L: a probe of new physics and tests using the Bell-Steinberger relation. Phys. Rev. **D91**(1), 015004 (2015). arXiv:1409.2288 [hep-ph]. https://doi.org/10.1103/PhysRevD.91.015004

73. A. Sommerfeld, Über die beugung und bremsung der elektronen. Annalen der Physik **403**, 257 (1931)

74. E. Fermi, An attempt of a theory of beta radiation. 1. Zeitschrift für Physik **88**, 161 (1934)

75. E. Gabrielli, M. Heikinheimo, B. Mele, M. Raidal, Dark photons and resonant monophoton signatures in Higgs boson decays at the LHC. Phys. Rev. **D90**(5), 055032 (2014). arXiv:1405.5196 [hep-ph]. https://doi.org/10.1103/PhysRevD.90.055032

76. S. Biswas, E. Gabrielli, M. Heikinheimo, B. Mele, Higgs-boson production in association with a dark photon in $e^+e^?$ collisions. JHEP **06**, 102 (2015). https://doi.org/10.1007/JHEP06(2015)102. arXiv:1503.05836 [hep-ph]

77. S. Biswas, E. Gabrielli, M. Heikinheimo, B. Mele, Dark-photon searches via ZH production at e^+e^- colliders. Phys. Rev. **D96**(5), 055012 (2017). arXiv:1703.00402 [hep-ph]. https://doi.org/10.1103/PhysRevD.96.055012

78. M. Fabbrichesi, E. Gabrielli, B. Mele, Z Boson decay into light and darkness. Phys. Rev. Lett. **120**, (17), 171803 (2018). arXiv:1712.05412 [hep-ph]. https://doi.org/10.1103/PhysRevLett.120.171803

79. M. Cobal, C. De Dominicis, M. Fabbrichesi, E. Gabrielli, J. Magro, B. Mele, G. Panizzo, Z-boson decays into an invisible dark photon at the LHC, HL-LHC and future lepton colliders. Phys. Rev. D **102**(3), 035027 (2020). arXiv:2006.15945 [hep-ph]. https://doi.org/10.1103/PhysRevD.102.035027

80. R. Barbieri, C. Braggio, G. Carugno, C.S. Gallo, A. Lombardi, A. Ortolan, R. Pengo, G. Ruoso, C.C. Speake, Searching for galactic Axions through magnetized media: the QUAX proposal. Phys. Dark Univ. **15**, 135–141 (2017). arXiv:1606.02201 [hep-ph]. https://doi.org/10.1016/j.dark.2017.01.003

81. S. Chigusa, T. Moroi, K. Nakayama, Detecting Light Boson Dark Matter through Conversion into Magnon. arXiv:2001.10666 [hep-ph]

82. D. Croon, A.E. Nelson, C. Sun, D.G.E. Walker, Z.-Z. Xianyu, Hidden-Sector spectroscopy with gravitational waves from binary neutron stars. Astrophys. J. **858**(1), L2 (2018). arXiv:1711.02096 [hep-ph]. https://doi.org/10.3847/2041-8213/aabe76

83. S. Alexander, E. McDonough, R. Sims, N. Yunes, Hidden-Sector modifications to gravitational waves from binary inspirals. Class. Quant. Grav. **35**(23), 235012 (2018). arXiv:1808.05286 [gr-qc]. https://doi.org/10.1088/1361-6382/aaeb5c

84. J. Kopp, R. Laha, T. Opferkuch, W. Shepherd, Cuckoo's eggs in neutron stars: can LIGO hear chirps from the dark sector? JHEP **11**, 096 (2018). https://doi.org/10.1007/JHEP11(2018)096. arXiv:1807.02527 [hep-ph]

85. M. Fabbrichesi, A. Urbano, Charged neutron stars and observational tests of a dark force weaker than gravity. arXiv:1902.07914 [hep-ph]

Chapter 3
Phenomenology of the Massive Dark Photon

The phenomenology of the massive dark photon is discussed in terms of its interaction with the SM particles, as given by Eq. (1.5):

$$\mathcal{L} = -\varepsilon e J^{\mu} A'_{\mu} , \tag{3.1}$$

where J^{μ} is the electromagnetic current. The strength of this interaction is modulated by the parameter ε. The parameter space for the experimental searches is given by the mass of the dark photon $m_{A'}$ and the mixing parameter ε.

3.1 Production, Decays and Detection

Because the current in Eq. (3.1) is the same as the usual electromagnetic current, dark photons A' can be produced like ordinary photons. The main production mechanisms are (Fig. 3.1):

- *Bremsstrahlung*: The incoming electron scatters off the target nuclei (Z), goes off-shell and can thus emit the dark photon: $e^- Z \rightarrow e^- Z A'$;
- Annihilation: An electron-positron pair annihilates into an ordinary and a dark photon: $e^- e^+ \rightarrow \gamma A'$
- Meson decays: A meson M (it being a π^0 η, or a K or a D) decays as $M \rightarrow \gamma A'$;
- Drell-Yan: A quark-antiquark pair annihilates into the dark photon, which then decays into a lepton pair (or hadrons): $q\bar{q} \rightarrow A'(\rightarrow \ell^+\ell^-$ or $h^+h^-)$.

Different experiments use different production mechanisms and, sometime, more than one simultaneously.

© The Author(s), under exclusive license to Springer Nature Switzerland AG 2021
M. Fabbrichesi et al., *The Physics of the Dark Photon*,
SpringerBriefs in Physics, https://doi.org/10.1007/978-3-030-62519-1_3

Fig. 3.1 Production of dark
photons: *Bremsstrahlung*,
Annihilation, Meson decay
and Drell-Yan

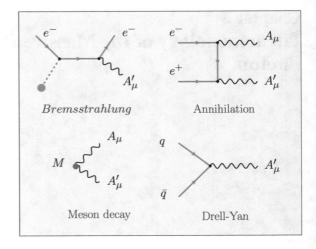

Detection of A' is based on its decays modes. The decay width of the massive
dark photon A' into SM leptons ℓ is

$$\Gamma(A' \to \ell^+\ell^-) = \frac{1}{3}\alpha\,\varepsilon^2 m_{A'}\sqrt{1 - \frac{4m_\ell^2}{m_{A'}^2}}\left(1 + \frac{2m_\ell^2}{m_{A'}^2}\right), \qquad (3.2)$$

which is only open for $m_{A'} > 2m_e$. Similarly, the width into hadrons is

$$\Gamma(A' \to \text{hadrons}) = \frac{1}{3}\alpha\varepsilon^2 m_{A'}\sqrt{1 - \frac{4m_\mu^2}{m_{A'}^2}}\left(1 + \frac{2m_\mu^2}{m_{A'}^2}\right)R, \qquad (3.3)$$

where $R \equiv \sigma_{e^+e^- \to \text{had}}/\sigma_{e^+e^- \to \mu^+\mu^-}$.

Since all visible widths are proportional to ε, the branching ratios are independent
of it.

At accelerator-based experiments, several approaches can be pursued to search
for dark photons depending on the characteristics of the available beam line and the
detector. These can be summarized as follows:

– Detection of visible final states: dark photons with masses above $\sim$1 MeV can
  decay to visible final states. The detection of visible final state is a technique
  mostly used in beam-dump and collider experiments, where typical signatures are
  expected to show up as narrow resonances over an irreducible background. Collider
  experiments are typically sensitive to larger values of ε ($\varepsilon > 10^{-3}$) than beam dump
  experiments which typically cover couplings below 10^{-3}. The use of this technique
  requires high luminosity colliders or large fluxes of protons/electrons on a dump
  because the dark photon detectable rate is proportional to the fourth power of the
  coupling involved, ε^4, and so very suppressed for very feeble couplings.

The smallness of the couplings implies that the dark photons are also very long-lived (up to 0.1 s) compared to the bulk of the SM particles. Hence: The decays to SM particles can be optimally detected using experiments with long decay volumes followed by spectrometers with excellent tracking systems and particle identification capabilities.

- Missing momentum/energy techniques: invisible decay of dark photons can be detected in fixed-target reactions as, for example, $e^- Z \to e^- Z A'$ (Z being the nuclei atomic number) with $A' \to \chi \overline{\chi}$ and χ being a putative dark matter particle, by measuring the missing momentum or missing energy carried away from the escaping invisible particle or particles. The main challenge for this approach is the very high background rejection that must be achieved, which relies heavily on the detector being hermetically closed and, in some cases, on the exact knowledge of the initial and final state kinematics.

  These techniques guarantee an intrinsic better sensitivity for the same luminosity than the technique based on the detection of dark photons decaying to visible final states, as it is independent of the probability of decays and therefore scales only as the SM-dark photon coupling squared, ε^2.

- Missing mass technique:

  This technique is mostly used to detect invisible particles (as DM candidates or particles with very long lifetimes) in reactions with a well-known initial state, as for example, at $e^+ e^-$ collider experiments using the process $e^+ e^- \to A'\gamma$, where A' is on shell, using the single photon trigger.

  Characteristic signature is the presence of narrow resonances emerging over a smooth background in the distribution of the missing mass.

  It requires detectors with very good hermeticity that allow to detect all the other particles in the final state. Characteristic signature of this reaction is the presence of a narrow resonance emerging over a smooth background in the distribution of the missing mass. The main limitation of this technique is the required knowledge of the background arising from processes in which particles in the final state escape the apparatus without being detected.

3.2 Visible and Invisible Massive Dark Photon

In collecting the limits on the parameters of massive dark photon is important to distinguish two cases accordingly on whether its mass is smaller or larger than twice the mass of the electron, the lightest charged SM fermion.

The dark photon is *visible* if its mass is $M_{A'} > 2m_e \simeq 1$ MeV because it can decay into SM charged states which leave a signature in the detectors (Fig. 3.2, top). We discuss the limits on the visible dark photon in Sect. 3.3.1.

In the same regime for which $M_{A'} > 1$ MeV, however, the massive dark photon could also decay into dark sector states if their masses are light enough. In this case we have a non-vanishing branching ratio into invisible final states. The invisible decay into these states of the dark sector χ in given by (Fig. 3.2, bottom).

Fig. 3.2 Decay of the
massive dark photon into
visible (SM leptons or
hadrons) and invisible (DM)
modes

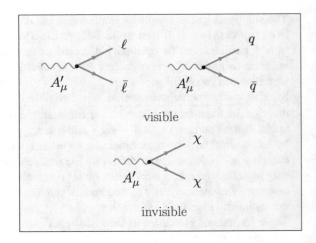

$$\Gamma(A' \to \chi\bar{\chi}) = \frac{1}{3}\alpha_D \, m_{A'} \sqrt{1 - \frac{4m_\chi^2}{m_{A'}^2}} \left(1 + \frac{2m_\chi^2}{m_{A'}^2}\right). \qquad (3.4)$$

Dark photons decays into this invisible channel if $m_{A'} > 2m_\chi$; this channel dominates
if $\alpha_D \gg \alpha\varepsilon^2$.

Most of the experimental searches with dark photon in visible decays assume that
the dark-sector states are not kinematically accessible and the dark photon is *visible*
only through its decay into SM states. The limits need to be re-modulated if the
branching ratio into invisible states is numerically significant or even dominant. We
discuss this case in Sect. 3.3.2 below.

If the mass of the dark photon is less than 1 MeV, it cannot decay in any known SM
charged fermion and its decay is therefore completely *invisible*. The experimental
searches for dark photon into invisible final states are based on the energy losses that
the production of dark photons, independently of his being stable or decaying into
dark fermions, implies on astrophysical objects like stars or in signals released in
direct detection dark matter experiments. The experimental limits in the case of the
invisible dark photon are discussed in Sect. 3.3.3 below.

3.3 Limits on the Parameters ε and $m_{A'}$

As discussed, the space of the parameters (the mixing ε and the mass $m_{A'}$ of the dark
photon) is best spanned in two regions according on whether the mass $m_{A'}$ is larger
or smaller than twice the mass of the electron: Roughly 1 MeV.

3.3.1 Constraints for $m_{A'} > 1\,MeV$ with A' Decays to Visible Final States

Two kinds of experiments provide the existing limits on the visible massive dark photon in the region of $m_{A'} > 1$ MeV: experiments at colliders and at fixed-target or beam dumps. In both cases the experiments search for resonances over a smooth background, with a vertex prompt or slightly displaced with respect to the beam interaction point in case of collider, or highly displaced in case of beam dump based experiments. The two categories are highly complementary, being the first category mostly sensitive to relatively large values of the mixing parameter ε, ($\varepsilon > 10^{-3}$) and the dark photon mass (up to several tens of GeV for pp collider experiments), while the second is sensitive to relatively small values ($10^{-7} \lesssim \varepsilon \lesssim 10^{-3}$) in the low mass range, $m_{A'}$ less than few GeV.

- Experiments at colliders. These experiments search for resonances in the invariant mass distribution of e^+e^-, $\mu^+\mu^-$ pairs. Different dark-photon production mechanisms are used in the different experiments: meson decays ($\pi^0 \rightarrow \gamma A'$, NA48/2 [9]), Bremsstrahlung ($e^- Z \rightarrow e^- Z A'$, A1 [1]), annihilation ($e^+e^- \rightarrow \gamma A'$, BaBar [4]), and all these processes in different searches at KLOE [5–8]. In a proton-proton (pp) collider the dark photon is produced via the $\gamma - A'$ mixing in all the processes where an off-shell photon γ^* with mass $m(\gamma^*)$ is produced: meson decays, Bremsstrahlung, and Drell-Yan production. LHCb [2, 37] has performed a search for dark photon decaying in $\mu^+\mu^-$ final states using 1.6 fb^{-1} of data collected at the LHC pp collisions at 13 TeV centre-of-mass energy. CMS [3] has performed the same search using 137 fb^{-1} of fully reconstructed data and 96.6 fb^{-1} of data collected with a reduced trigger information.
  Figure 3.3 shows the existing limits for NA48/2, A1, LHCb, and BaBar; only one set of limits from KLOE is shown since the others have been superseded by the limits from BaBar.
- Beam-dump experiments. These experiments use the collisions of an electron or proton beam with a fixed-target or a dump to generate the dark photon via Bremsstrahlung (electron and proton beams), meson production and QCD processes (proton beams only). The products of the collisions are mostly absorbed in the dump and the dark photon is searched for as a displaced vertex with two opposite charged tracks in the decay volume of the experiment.
  Figure 3.3 shows the limits from experiments at extracted electron beams (E141 [11] and E137 [12–14] at SLAC, E774 [10] at Fermilab) and at extracted proton beams from CHARM at CERN ([17] based on CHARM data [38]).

In addition, bounds on energy losses in supernovae provide further limits in the region of small masses. These limits where discussed in [39, 40] and updated in [18, 41] by including the effect of finite temperature and plasma density.

Also the electron magnetic moment, with its very precise experimental determination, can be used to set an indirect limit [19]. These limits are included in Fig. 3.3.

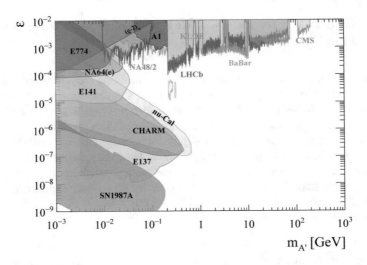

Fig. 3.3 Existing limits on the massive dark photon for $m_{A'} > 1$ MeV from di-lepton searches at experiments at collider/fixed target (A1 [1], LHCb [2], CMS [3], BaBar [4], KLOE [5–8], and NA48/2 [9]) and old beam dump: E774 [10], E141 [11], E137 [12–14]), ν-Cal [15, 16], and CHARM (from [17]. Bounds from supernovae [18] and $(g − 2)_e$ [19] are also included

Recent constraints from ATLAS [42, 43] and CMS [44] would nominally cover the interesting region around 1 GeV for ε between 10^{-6} and 10^{-2} but unfortunately they have been framed within a restrictive model and are not on the same footing that the limits included in Fig. 3.3.

Additional limits (not included in Fig. 3.3) from cosmology (in the cosmic microwave background and nucleosynthesis) exist in the very dark region of very small $\varepsilon < 10^{-10}$ [45].

Looking at Fig. 3.3, it is clear that it would be desirable to first close the gap between the beam-dump and the collider based experiments in the region between tens of MeV up to 1 GeV in the dark photon mass, and then extend the limits for larger masses. Both of these goals could be achieved through a series of experiments summarized here below whose sensitivity is shown in Fig. 3.4 as colored curves.

- *Belle-II at SuperKEKB* will search for visible dark photon decays $A' \rightarrow e^+e^-$, $\mu^+\mu^-$ where A' is produced in the process $e^+e^- \rightarrow A'\gamma$. The projections shown in Fig. 3.4 is based on 50 ab^{-1} of integrated luminosity [20].
- *LHCb upgrade (phase I and phase II) at the LHC*: LHCb phase I will search for dark photon in visible final states both using the inclusive di-muon production [21] and the $D^{*0} \rightarrow D^0 e^+e^-$ decays [22]. The projections are based on 15 fb^{-1}, 3 years data taking with 5 fb^{-1}/year with an upgraded detector after the LHC Long Shutdown 2. This can be further improved with a possible Phase II upgraded detector [46] collecting up to 300 fb^{-1} of integrated luminosity after Long Shutdown 4.

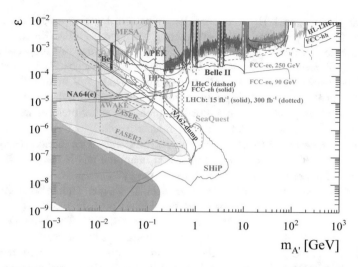

Fig. 3.4 Colored curves are projections for existing and proposed experiments on the massive dark photon for $m_{A'} > 1$ MeV: Belle-II [20] at SuperKEKb; LHCb upgrade [21, 22] at the LHC; NA62 in dump mode [23] and NA64(e)[++] [24] at the SPS; FASER and FASER2 [25] at the LHC; SeaQuest [26] at Fermilab; HPS [27] at JLAB; an NA64-like experiment at AWAKE [28], and an experiment dedicated to dark photon searches at MESA [29, 30]. For masses above 10 GeV projections obtained for ATLAS/CMS during the high luminosity phase of the LHC (HL-LHC [31]) and for experiments running at a future FCC-ee [32], LHeC/FCC-eh [33], and FCC-hh [31] are also shown. The vertical red line shows the allowed range of couplings of a new gauge boson X to electrons that could explain the ^{8}Be anomaly [34, 35]. The existing limits are shown as gray areas. The bottom plot is revised from [36]

- *NA62[++] or NA62 in dump mode at the SPS, CERN,* will search for a multitude of feebly-interacting particles, including dark photon, decaying into visible final states and possibly emerging from the interactions of 400 GeV proton beam with a dump. NA62 aims to collect approximately 10^{18} protons-on-target in 2021–2024 [23].
- *NA64(e)[++] at the SPS, CERN,* is the upgrade of the existing NA64(e) experiment. It aims to collect about 5×10^{12} electrons-on-target after the CERN Long Shutdown 2 [24].
- *NA64-like experiment at AWAKE, CERN*: progress in the coming years in proton-driven plasma wake-field acceleration of electrons at the AWAKE facility at CERN could allow an NA64-like experiment be served by a high-intensity high energy primary electron beam for search for dark photons in visible final states [28]. The sensitivity plot has been obtained assuming $\sim 10^{16}$ electrons-on-target with an energy of 50 GeV.
- *FASER and FASER2 at the LHC, CERN:* FASER [25] is being installed in a service tunnel of the LHC located along the beam collision axis, 480 m downstream from the ATLAS interaction point. At this location, FASER (and possibly its larger successor FASER2) will enhance the LHC discovery potential by providing sensitivity

to dark photons, dark Higgs bosons, heavy neutral leptons, axion-like particles, and many other proposed feebly-interacting particles [47]. FASER and FASER2 aim to collect 150 fb^{-1} and 3000 fb^{-1} of integrated luminosity, respectively.

- *HPS at Jefferson Laboratory (JLab)*: The HPS experiment [27], proposed at an electron beam-dump at CEBAF electron beam (2.2–6.6 GeV, up to 500 nA), search for visible ($A' \to e^+e^-$) dark photon (prompt and displaced) decays produced via *Bremsstrahlung* production in a thin W target. The experiment makes use of the 200 nA electron beam available in Hall-B at Jefferson Lab.

- *SeaQuest at Fermilab (FNAL)*: will search for visible dark photon decays $A' \to e^+e^-$ at the 120 GeV main injector proton beamline at FNAL [26]. It plans to accumulate approximately 10^{18} protons-on-target by 2024.

- *MAGIX or Beam Dump Experiment at MESA, Mainz*: The MESA accelerator is a *continuous wave* linac that will be able to provide an electron beam of $E_{max} = 155$ MeV energy and up to 1 mA current [29]. The MAGIX detector is a twin arm dipole spectrometer placed around a gas target and will search for search for visible ($A' \to e^+e^-$) dark photon (prompt and displaced) decays produced via *Bremsstrahlung* production [30]. The possibility of a beam dump setup experiment is also under study. Timeline: targeted operations in 2021–2022 and 2 years of data taking.

- *Experiments at a future e^+e^- circular collider, FCC-ee*: a powerful technique to be exploited at experiments running at a future e^+e^- circular collider is the radiative return, $e^+e^- \to A'\gamma, A' \to \mu^+\mu^-$. The results obtained in [32] have been rescaled to the integrated luminosities of 150 fb^{-1} at $\sqrt{s} = 90$ GeV and 5 ab^{-1} at $\sqrt{s} = 250$ GeV, as in [48].

- *ATLAS/CMS at the high-luminosity phase at the LHC and at a future pp circular collider*: at pp colliders the dark photon can be produced via a Drell-Yan process, $pp \to A' \to e^+e^-, \mu^+\mu^-$. The physics reach of ATLAS/CMS like experiments have been computed for $\sqrt{s} = 14$ TeV and 3 ab^{-1} and $\sqrt{s} = 100$ TeV, 3 ab^{-1} [31].

- *ATLAS/CMS at a possible LHeC collider in the LHC tunnel and a future FCC-eh circular collider*: At the LHeC (FCC-eh) a 7 TeV (50 TeV) a proton beam collides with a 60 GeV electron beam achieving a center-of-mass energy of 1.3 TeV (3.5 TeV) and a total integrate luminosity of 1 ab^{-1} (3 ab^{-1}). At eh colliders the main production process for the dark photon is the deep inelastic scattering $e^- +$ parton $\to e^-$ parton A', with $A' \to$ charged fermions [33].

3.3.2 Constraints for $m_{A'} > 1$ MeV with A' Decays to Invisible Final States

Different constraints apply in the case of massive dark photon going into invisible final states in the mass region $m_{A'} > 1$ MeV. In this case techniques like missing momentum, missing energy, and missing mass are used in order to identify a possible massive dark photon decaying into invisible final states.

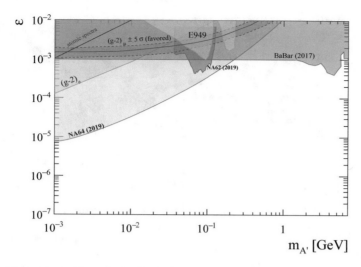

Fig. 3.5 Existing limits for a massive dark photon going to invisible final states ($\alpha_D >> \alpha\varepsilon^2$). Existing limits from Kaon decay experiments (E787 [49], E949 [50], NA62 [51]), BaBar [52], and NA64(e) [53]. The constraints from $(g-2)_\mu$ [54] and $(g-2)_e$ are also shown

The most stringent bounds come from BaBar [52] and the electron beam dump NA64(e) experiment at CERN [53] which recently superseded the results from Kaon experiments (E787 [49] and E949 [50] at BNL, NA62 [51] at CERN). The existing bounds are depicted in the top plot of Fig. 3.5 as colored areas. These limits overlap with the exclusion regions defined by the dark photon decays into visible final states for masses $m_{A'} > 1$ GeV and complement them in the range of masses 10 MeV $\lesssim m_{A'} \lesssim 1$ GeV and kinetic mixing strength $10^{-5} \lesssim \varepsilon \lesssim 10^{-3}$, where the searches of dark photon into visible decays are typically weaker.

Sensitivities of existing or proposed experiments are shown in Fig. 3.6 as colored lines. These include:

– *NA64(e)$^{++}$* with 5×10^{12} electrons-on-target will search $A' \rightarrow$ invisible final states with a missing energy technique using a secondary electron beam at ~ 100 GeV at the CERN SPS [24].

– *Belle II* will search for dark photons in the process $e^+e^- \rightarrow A'$ and $A' \rightarrow$ invisible [20]. Projections are based on 20 fb^{-1} of integrated luminosity.

– *KLEVER*, proposed at the SPS, could search for dark photons in invisible final states as a by-product of the analysis of the $K_L \rightarrow \pi^0\nu\bar{\nu}$ rare decay, pushing further the investigation performed by traditional Kaon experiments in the mass region between 100 and 200 MeV [55].

– *PADME* [56] will search for $A' \rightarrow$ invisible final states using the missing momentum technique at the Beam Test Facility (BTF) at Laboratori Nazionali di Frascati (INFN). It will use a 550 MeV positron beam on a diamond target. A first com-

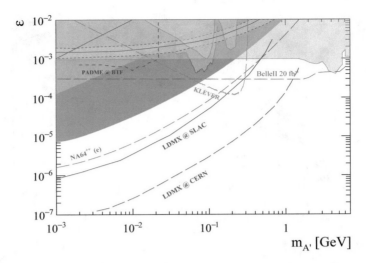

Fig. 3.6 Future sensitivities for proposed experiments for a massive dark photon going to invisible final states ($\alpha_D >> \alpha\varepsilon^2$). Future sensitivities for NA64(e)$^{++}$ [24], Belle II [20], KLEVER [55], PADME [56], LDMX@SLAC [57, 58], and LDMX@CERN [57, 59]. The sensitivity curves for LDMX@SLAC and LDMX@CERN assume 10^{14} electrons-on-target and $E_{beam} = 4$ GeV and 10^{16} electrons-on-target and $E_{beam} = 16$ GeV, respectively. The bottom plot is revised from [36]. See text for details

missioning run was performed in late 2018 and early 2019 to assess the detector performance and beam line quality. A physics data taking to collect 5×10^{12} positrons on target is expected in the second part of 2020.

3.3.3 Constraints for $m_{A'} < 1$ MeV

Strong constraints exist for the invisible massive dark photon in the region $m_{A'} < 1$ MeV. They come from different sources:

- Atomic and nuclear experiments: These experiments aim to detect modifications of the Coulomb force (as discussed in [85]) due to the dark photon. Corrections in Rydberg atoms, Lamb shift and hyperfine splitting in atomic hydrogen have been translated into bounds on the massive dark photon mixing parameter [70]. The results of the TEXONO neutrino experiment [86] have been interpreted in terms of dark photon parameters in [69];
- Axion-like particles and helioscopes: Experiments of light shining through a wall (LSW) for axions and axion-like particles can be adapted to the dark photon and limits can accordingly be estimated [64]. The same phenomenon has been used in the experiment CROWS [65] at CERN. The CAST result, on the flux of axion-like particles from the Sun (Helioscope), can be translated [66] into a bound on the

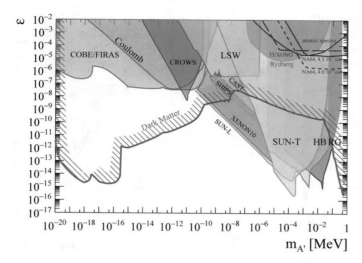

Fig. 3.7 Current limits on massive dark photon for $m_{A'} < 1$ MeV. Bounds from cosmology (COBE/FIRES [60–63]), light through a wall (LSW) [64], CROWS [65], CAST [66], XENON10 [67], SHIPS [68], TEXONO [69], atomic experiments (Coulomb, Rydberg and atomic spectra [70]) and astrophysics: Solar lifetime (SUN-T and SUN-L), red giants (RG), horizontal branches (HB) [41, 71, 72]. Additional limits under the assumption that the dark photon is the dark matter: The curve "Dark Matter" includes the combination of the constraints from the references discussed in the main text

massive dark photon parameters. The same is true for XENON10, whose data set provides further limits [67] and the results from the experiment SHIPS [68];

- Astrophysics: The non-observation of anomalous energy transport (by the mechanism discussed in Sect. 2.1) in stars on the horizontal branch (HB), red giants (RG) and the Sun (SUN-T and SUN-L) imposes severe constraints on the mixing parameter of the massive dark photon. Mixing effects are important in these processes for both the longitudinal (L) and transverse (T) modes and one must use thermal field theory [41, 71, 72]. The dark photon partakes of the plasmon modes (see Sect. 5.3) in an effective mixing with the ordinary photon proportional to its mass (and vanishing as it goes to zero).
- Cosmology: The oscillation between the ordinary and the massive dark photon $\gamma \to A'$ induces deviations on the black body spectrum (as measured by COBE/FIRAS [87]) in the cosmic microwave background. This effect depends on the effective plasma mass of the dark photon and it is enhanced when this mass is equal to $m_{A'}$. The bound depicted in Fig. 3.7 follows the most recent evaluation [60–62]—which includes inhomogeneities in the plasma mass—for values $m_{A'} < 10^{-15}$ MeV, and [63, 88] for larger values.

Even stronger constraints can be derived under the assumption that the dark photon is itself dark matter. The combination (in order of increasing values of $m_{A'}$) of

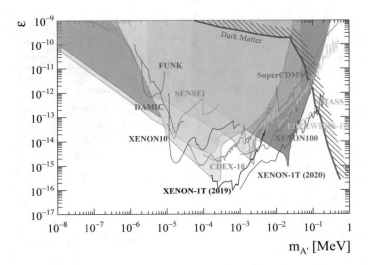

Fig. 3.8 Zoom in the range 10^{-8} MeV $\lesssim M_{A'} \lesssim$ 1 MeV and $10^{-17} \lesssim \varepsilon \lesssim 10^{-9}$ for the current limits on massive dark photon for $m_{A'} < 1$ MeV. Results from dark matter direct detection experiments (XENON10 and XENON100 ([73] based on XENON10 [75] and XENON100 [74] data), XENON1T (2019) [76]; XENON1T (2020) [77]; DAMIC [78]; SuperCDMS [79]; CDEX-10 [80]; EDELWEISS-III [81]; SENSEI [82]; XMASS [83]; FUNK [84])

– astrophysical bounds on dwarf galaxies [89]. These limits apply for values of $m_{A'} < 10^{-19}$ MeV and are not shown in Fig. 3.7 measurements of the temperature of the intergalactic medium at the epoch of He^{++} re-ionization in the presence of inhomogeneities [60–62],
– cold Galactic Center gas clouds heating rates [90];
– cosmic microwave background spectral distortions (μ and y-type) [62];
– energy deposition during the dark ages [63],
– the number of relativistic species ΔN^{eff} during big-bang nucleosynthesis and recombination [91] and
– the diffuse X-ray background [92]

yield a series of limits on the upper value of ε for different values of $M_{A'}$. These limits are depicted together by the curve labelled "Dark Matter" in Fig. 3.7.

In addition, there are limits from:

• Dark matter direct detection experiments: These experiments are part of the ongoing search for dark matter through its direct detection. Data from XENON10/XENON100 ([73] based on XENON10 [75] and XENON100 [74]), XENON1T [76], DAMIC [78], SuperCDMS [79], CDEX-10 [80], EDELWEISS-III [81], SENSEI [82], XMASS [83] and FUNK [84] can be used to constrain the massive dark photon parameters;
• Haloscopes: Searches with microwave cavities for relic axion converting to photons [93] can be translated into limits (not shown in Fig. 3.7 on the dark photon

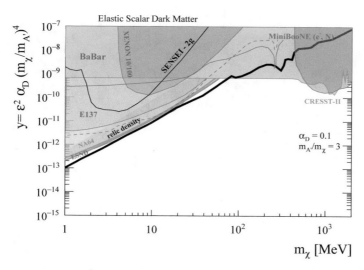

Fig. 3.9 Existing limits for existing experiments for massive dark photon for $m_{A'} > 1$ MeV in the plane of the yield variable y as a function of dark matter mass m_χ for an elastic scalar dark matter particle. Limits from BaBar [52], NA64(e) [53], reinterpretation of the data from E137 [13] and LSND [94]; result from MiniBooNE [95]; interpretation in the dark photon framework of data from CRESST-II [96]

parameter ε to be less than $10^{-13} - 10^{-15}$ in the range around $10^{-11} - 10^{-12}$ MeV [91].

The limits from dark-matter direct detection are shown in Fig. 3.8.

Some of the limits on the right side of Fig. 3.7 are the continuation of the corresponding left side of the limits in Fig. 3.5. The two figures are back-to-back at $m_{A'} = 1$ MeV thus covering the full range of the dark-photon masses.

3.4 Limits on the Parameters y and m_χ

If the dark sector states into which the invisible dark photon decays are taken to be dark matter, there are new limits involving also the coupling strength α_d and the connection to the direct-detection searches for dark matter. As discussed in Sect. 1.3, the best way to plot the experimental limits in this case is in terms of the yield variable y, defined in Eq. (1.28), and the dark matter mass m_χ.

The corresponding limits strongly depend on the nature of the dark-matter state χ because the velocity dependence of the averaged cross sections. In the case of Dirac fermions, Planck data [102] rule out sub-GeV dark matter because of their too large annihilation rate at the cosmic microwave background epoch. For this reason,

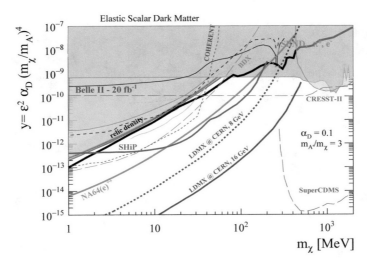

Fig. 3.10 Future sensitivities for proposed experiments for massive dark photon for $m_{A'} > 1$ MeV in the plane of the yield variable y as a function of dark matter mass m_χ for an elastic scalar dark matter particle. Projections for SHiP [97], BDX [98], SBND [99], LDMX@CERN [57, 59], SENSEI with a proposed 100 g detector operating at SNOLAB [100], and SuperCDMS at SNOLAB [101]. The plot is revised from [36]

pseudo-Dirac fermions and scalars, which have velocity suppressed annihilation cross sections, are usually studied.

The current bounds and future perspectives in the plane y versus dark matter mass are shown in Figs. 3.9 and 3.10 under the hypothesis that the dark matter is a scalar particle and for a specific choice of α_d ($\alpha_d = 0.1$) and the ratio between the mediator and the dark matter masses ($m_{A'}/m_\chi = 3$). In these plots, the lower limit for the thermal relic density is also shown, under that hypothesis that a single dark-matter candidate is responsible for the whole dark-matter abundance. It is worth noting that results from accelerator-based experiments are largely independent of the assumptions on a specific dark matter nature as dark matter at accelerators is produced in relativistic regime and the strength of the interactions with light mediators and SM particles is only fixed by thermal freeze-out.

Current bounds come from the same experiments using missing energy/missing momentum techniques contributing to the $\{\varepsilon, m_{A'}\}$ sensitivity plot (BaBar and NA64(e)) with the addition of the re-interpretation of data from old neutrino experiments (E137 [13] and LSND [94]) and results from current neutrino experiments (MiniBooNE [95]) exploiting dark matter scattering on nucleons and/or electrons. Bounds can also be derived by using a superfluid He-4 detector, as shown in [103], but they lie at the margin of the range included in Fig. 3.9.

In all the bounds shown for electron beam-dump or missing energy experiments, we are neglecting the contribution of secondary positrons in dark photon production through annihilation on atomic electrons, as for example studied in [14].

Future initiatives that could explore a still uncovered parameter space in the plane $\{y, m_\chi\}$ for dark matter masses below 1 GeV are all those that have sensitivity in the plane $\{\varepsilon, m_{A'}\}$ and, in addition, accelerator-based and dark matter direct detection experiments exploiting dark matter scattering against the nucleons and/or electrons. Accelerator-based experiments are SHiP at CERN [97], and BDX at JLab [98] and SBND [99] at FNAL as explained below.

- *BDX at JLAB* the Beam Dump eXperiment (BDX) [98] is aiming to detect light dark matter χ produced in the interaction of an intense (100 μA) 10 GeV electron beam with a dump. The experiment is sensitive to elastic dark matter scattering $e^-\chi \to e^-\chi$ in the detector after production in $e^-Z \to e^-ZA'(A' \to \chi\chi)$.
- *SBND* is planned to be installed at the 8 GeV proton Booster Neutrino Beamline at FNAL about 470 m downstream of the beam dump [99]. The dark matter beam is primarily produced via pion decays out of collisions from the primary proton beam, and identified via dark-matter-nucleon or dark-matter-electron elastic scattering in a LAr-based detector. SBND is expected to improve upon MiniBooNE by more than an order of magnitude with 6×10^{20} protons-on-target.

Also dark matter direct-detection experiments with sensitivity below 1 GeV mass contribute to this plot. These are:

- *SENSEI* is a direct detection experiment [104] that will be able to explore dark matter candidates with masses in the 1 eV and few GeV range, by detecting the signal released in dark-matter-electron scattering interactions in a fully depleted silicon CCD. A 2-gram detector is already operating in the NUMI access tunnel [105]. A larger project (100 grams) can be deployed at SNOLAB if funding is obtained [100].
- *CRESST-II* [96] uses cryogenic detectors to search for nuclear recoil events induced by elastic scattering of dark-matter particles in $CaWO_4$ crystals. Because of its low-energy threshold, the sensitivity to dark matter was extended in the sub-GeV region. Current bounds are derived from a dataset corresponding to 52 kg live days.
- *super-CDMS* [101] at SNOLAB (Canada)Start mid-2021, and uses 30 kg of Germanium and Silicium detectors.

References

1. H. Merkel et al., Search at the mainz microtron for light massive Gauge Bosons relevant for the Muon g-2 Anomaly. Phys. Rev. Lett. **112**(22), 221802 (2014). arXiv:1404.5502 [hep-ex]. https://doi.org/10.1103/PhysRevLett.112.221802
2. LHCb Collaboration, R. Aaij et al., Search for $A' \to \mu^+\mu^-$ Decays. Phys. Rev. Lett. **124**(4), 041801 (2020). arXiv:1910.06926 [hep-ex]. https://doi.org/10.1103/PhysRevLett.124.041801
3. CMS Collaboration, A.M. Sirunyan et al., Search for a narrow resonance decaying to a pair of muons in proton-proton collisions at 13 TeV

4. BaBar Collaboration, J.P. Lees et al., Search for a dark photon in e^+e^- collisions at BaBar. Phys. Rev. Lett. **113**(20), 201801 (2014). arXiv:1406.2980 [hep-ex]. https://doi.org/10.1103/PhysRevLett.113.201801

5. KLOE-2 Collaboration, F. Archilli et al., Search for a vector gauge boson in ϕ meson decays with the KLOE detector. Phys. Lett. **B706**, 251–255 (2012). arXiv:1110.0411 [hep-ex]. https://doi.org/10.1016/j.physletb.2011.11.033

6. KLOE-2 Collaboration, D. Babusci et al., Limit on the production of a light vector gauge boson in phi meson decays with the KLOE detector. Phys. Lett. **B720**, 111–115 (2013). arXiv:1210.3927 [hep-ex]. https://doi.org/10.1016/j.physletb.2013.01.067

7. KLOE-2 Collaboration, D. Babusci et al., Search for light vector boson production in $e^+e^- \to \mu^+\mu^-\gamma$ interactions with the KLOE experiment. Phys. Lett. **B736**, 459–464 (2014). arXiv:1404.7772 [hep-ex]. https://doi.org/10.1016/j.physletb.2014.08.005

8. KLOE-2 Collaboration, A. Anastasi et al., Limit on the production of a new vector boson in $e^+e^- \to U\gamma$, $U \to \pi^+\pi^-$ with the KLOE experiment. Phys. Lett. **B757**, 356–361 (2016). arXiv:1603.06086 [hep-ex]. https://doi.org/10.1016/j.physletb.2016.04.019

9. NA48/2 Collaboration, J.R. Batley et al., Search for the dark photon in π^0 decays. Phys. Lett. **B746**, 178–185 (2015). arXiv:1504.00607 [hep-ex]. https://doi.org/10.1016/j.physletb.2015.04.068

10. A. Bross, M. Crisler, S.H. Pordes, J. Volk, S. Errede, J. Wrbanek, A search for Shortlived particles produced in an electron beam dump. Phys. Rev. Lett. **67**, 2942–2945 (1991). https://doi.org/10.1103/PhysRevLett.67.2942

11. E.M. Riordan et al., A search for short lived Axions in an electron beam dump experiment. Phys. Rev. Lett. **59**, 755 (1987). https://doi.org/10.1103/PhysRevLett.59.755

12. J.D. Bjorken, S. Ecklund, W.R. Nelson, A. Abashian, C. Church, B. Lu, L.W. Mo, T.A. Nunamaker, P. Rassmann, Search for neutral metastable penetrating particles produced in the SLAC beam dump. Phys. Rev. D **38**, 3375 (1988). https://doi.org/10.1103/PhysRevD.38.3375

13. B. Batell, R. Essig, Z. Surujon, Strong constraints on Sub-GeV dark sectors from SLAC beam dump E137. Phys. Rev. Lett. **113**(17), 171802 (2014). arXiv:1406.2698 [hep-ph]. https://doi.org/10.1103/PhysRevLett.113.171802

14. L. Marsicano, M. Battaglieri, M. Bondi', C.R. Carvajal, A. Celentano, M. De Napoli, R. De Vita, E. Nardi, M. Raggi, P. Valente, Dark photon production through positron annihilation in beam-dump experiments. Phys. Rev. D **98**(1), 015031 (2018). arXiv:1802.03794 [hep-ex]. https://doi.org/10.1103/PhysRevD.98.015031

15. J. Blumlein, J. Brunner, New exclusion limits for dark gauge forces from beam-dump data. Phys. Lett. B **701**, 155–159 (2011). arXiv:1104.2747 [hep-ex]. https://doi.org/10.1016/j.physletb.2011.05.046

16. J. Blumlein, J. Brunner, New exclusion limits on dark gauge forces from proton bremsstrahlung in beam-dump data. Phys. Lett. **B731**, 320–326 (2014). arXiv:1311.3870 [hep-ph]. https://doi.org/10.1016/j.physletb.2014.02.029

17. S. Gninenko, Constraints on sub-GeV hidden sector gauge bosons from a search for heavy neutrino decays. Phys. Lett. B **713**, 244–248 (2012). arXiv:1204.3583 [hep-ph]. https://doi.org/10.1016/j.physletb.2012.06.002

18. J.H. Chang, R. Essig, S.D. McDermott, Revisiting supernova 1987A constraints on dark photons. JHEP **01**(107) (2017). https://doi.org/10.1007/JHEP01(2017)107. arXiv:1611.03864 [hep-ph]

19. M. Pospelov, Secluded U(1) below the weak scale. Phys. Rev. D **80**, 095002 (2009). https://doi.org/10.1103/PhysRevD.80.095002. arXiv:0811.1030 [hep-ph]

20. Belle-II Collaboration, W. Altmannshofer et al., The Belle II physics book. PTEP **2019**(12), 123C01 (2019). https://doi.org/10.1093/ptep/ptz106, https://doi.org/10.1093/ptep/ptaa008. arXiv:1808.10567 [hep-ex]. [Erratum: PTEP2020,no.2,029201(2020)]

21. P. Ilten, Y. Soreq, J. Thaler, M. Williams, W. Xue, Proposed inclusive dark photon search at LHCb. Phys. Rev. Lett. **116**(25), 251803 (2016). arXiv:1603.08926 [hep-ph]. https://doi.org/10.1103/PhysRevLett.116.251803

22. P. Ilten, J. Thaler, M. Williams, W. Xue, Dark photons from charm mesons at LHCb. Phys. Rev. **D92**(11), 115017 (2015). arXiv:1509.06765 [hep-ph]. https://doi.org/10.1103/PhysRevD.92. 115017

23. E. Cortina Gil et al., ADDENDUM I TO P326 Continuation of the physics programme of the NA62 experiment. https://cds.cern.ch/record/2691873?ln=en

24. D. Banerjee et al., Addendum to the NA64 proposal: search for $A' \to$ invisible and $X \to e^+e^-$ decays in 2021. http://cds.cern.ch/record/2300189?ln=en

25. J.L. Feng, I. Galon, F. Kling, S. Trojanowski, ForwArd search ExpeRiment at the LHC. Phys. Rev. **D97**(3), 035001 (2018). arXiv:1708.09389 [hep-ph]. https://doi.org/10.1103/PhysRevD. 97.035001

26. A. Berlin, S. Gori, P. Schuster, N. Toro, Dark sectors at the Fermilab SeaQuest experiment. Phys. Rev. **D98**(3), 035011 (2018). arXiv:1804.00661 [hep-ph]. https://doi.org/10. 1103/PhysRevD.98.035011

27. HPS Collaboration, P. Adrian et al., Search for a dark photon in electroproduced e^+e^- pairs with the Heavy Photon Search experiment at JLab. Phys. Rev. D **98**(9), 091101 (2018). arXiv:1807.11530 [hep-ex]. https://doi.org/10.1103/PhysRevD.98.091101

28. A. Caldwell et al., Particle physics applications of the AWAKE acceleration scheme. arXiv:1812.11164 [physics.acc-ph]

29. L. Doria, P. Achenbach, M. Christmann, A. Denig, H. Merkel, Dark matter at the intensity frontier: the new MESA electron accelerator facility, in *An Alpine LHC Physics Summit 2019 (ALPS 2019) Obergurgl, Austria*, 22–27 Apr 2019. arXiv:1908.07921 [hep-ex]

30. L. Doria, P. Achenbach, M. Christmann, A. Denig, P. Gulker, H. Merkel, Search for light dark matter with the MESA accelerator, in *13th Conference on the Intersections of Particle and Nuclear Physics (CIPANP 2018) Palm Springs, CA, USA*, 29 May–3 June 2018. arXiv:1809.07168 [hep-ex]

31. D. Curtin, R. Essig, S. Gori, J. Shelton, Illuminating dark photons with high-energy colliders. JHEP **02**, 157 (2015). https://doi.org/10.1007/JHEP02(2015)157. arXiv:1412.0018 [hep-ph]

32. M. Karliner, M. Low, J.L. Rosner, L.-T. Wang, Radiative return capabilities of a high-energy, high-luminosity e^+e^- collider. Phys. Rev. D **92**(3), 035010 (2015). https://doi.org/10.1103/ PhysRevD.92.035010. arXiv:1503.07209 [hep-ph]

33. M. D'Onofrio, O. Fischer, Z. S. Wang, Searching for dark photons at the LHeC and FCC-he. Phys. Rev. D **101**(1), 015020 (2020). arXiv:1909.02312 [hep-ph]. https://doi.org/10.1103/ PhysRevD.101.015020

34. J.L. Feng, B. Fornal, I. Galon, S. Gardner, J. Smolinsky, T.M. Tait, P. Tanedo, Protophobic fifth-force interpretation of the observed anomaly in ^{8}Be nuclear transitions. Phys. Rev. Lett. **117**(7), 071803 (2016). arXiv:1604.07411 [hep-ph]. https://doi.org/10.1103/PhysRevLett. 117.071803

35. J.L. Feng, B. Fornal, I. Galon, S. Gardner, J. Smolinsky, T.M.P. Tait, P. Tanedo, Particle physics models for the 17 MeV anomaly in beryllium nuclear decays. Phys. Rev. D **95**(3), 035017 (2017). arXiv:1608.03591 [hep-ph]. https://doi.org/10.1103/PhysRevD.95.035017

36. J. Beacham et al., Physics Beyond colliders at CERN: Beyond the standard model working group report. J. Phys. G **47**(1), 010501 (2020). arXiv:1901.09966 [hep-ex]. https://doi.org/ 10.1088/1361-6471/ab4cd2

37. LHCb Collaboration, R. Aaij et al. Search for dark photons produced in 13 TeV pp collisions. Phys. Rev. Lett. **120**(6), 061801 (2018). arXiv:1710.02867 [hep-ex]. https://doi.org/10.1103/ PhysRevLett.120.061801

38. CHARM Collaboration, F. Bergsma et al., Search for Axion like particle production in 400-GeV Proton—Copper interactions. Phys. Lett. **157B**, 458–462 (1985). https://doi.org/10. 1016/0370-2693(85)90400-9

39. J.B. Dent, F. Ferrer, L.M. Krauss, Constraints on Light Hidden Sector Gauge Bosons from Supernova Cooling. arXiv:1201.2683 [astro-ph.CO]

40. H.K. Dreiner, J.-F. Fortin, C. Hanhart, L. Ubaldi, Supernova constraints on MeV dark sectors from e^+e^- annihilations. Phys. Rev. **D89**(10), 105015 (2014). arXiv:1310.3826 [hep-ph]. https://doi.org/10.1103/PhysRevD.89.105015

41. E. Hardy, R. Lasenby, Stellar cooling bounds on new light particles: plasma mixing effects. JHEP **02**, 033 (2017). https://doi.org/10.1007/JHEP02(2017)033. arXiv:1611.05852 [hep-ph]

42. ATLAS Collaboration, G. Aad et al., Search for long-lived neutral particles decaying into lepton jets in proton-proton collisions at $\sqrt{s} = 8$ TeV with the ATLAS detector. JHEP **11**, 088 (2014). arXiv:1409.0746 [hep-ex]. https://doi.org/10.1007/JHEP11(2014)088

43. ATLAS Collaboration, G. Aad et al., A search for prompt lepton-jets in pp collisions at $\sqrt{s} = 8$ TeV with the ATLAS detector. JHEP **02**, 062 (2016). arXiv:1511.05542 [hep-ex]. https://doi.org/10.1007/JHEP02(2016)062

44. CMS Collaboration, V. Khachatryan et al., A search for pair production of new light bosons decaying into muons. Phys. Lett. **B752**, 146–168 (2016). arXiv:1506.00424 [hep-ex]. https://doi.org/10.1016/j.physletb.2015.10.067

45. A. Fradette, M. Pospelov, J. Pradler, A. Ritz, Cosmological constraints on very dark photons. Phys. Rev. **D90**(3), 035022 (2014). arXiv:1407.0993 [hep-ph]. https://doi.org/10.1103/PhysRevD.90.035022

46. LHCb Collaboration, R. Aaij et al., Physics case for an LHCb Upgrade II— Opportunities in flavour physics, and beyond, in the HL-LHC era. arXiv:1808.08865 [hep-ex]

47. FASER Collaboration, A. Ariga et al., FASER physics reach for long-lived particles. Phys. Rev. **D99**(9), 095011 (2019). arXiv:1811.12522 [hep-ph]. https://doi.org/10.1103/PhysRevD.99.095011

48. R.K. Ellis et al., Physics Briefing Book: Input for the European Strategy for Particle Physics Update 2020. arXiv:1910.11775 [hep-ex]

49. E787 Collaboration, S. Adler et al., Further evidence for the decay $K^+ \rightarrow \pi^+ \nu \bar{\nu}$. Phys. Rev. Lett. **88**, 041803 (2002). arXiv:hep-ex/0111091 [hep-ex]. https://doi.org/10.1103/PhysRevLett.88.041803

50. BNL-E949 Collaboration, A.V. Artamonov et al., Study of the decay $K^+ \rightarrow \pi^+ \nu \bar{\nu}$ in the momentum region $140 < P_\pi < 199$ MeV/c. Phys. Rev. **D79**, 092004 (2009). arXiv:0903.0030 [hep-ex]. https://doi.org/10.1103/PhysRevD.79.092004

51. NA62 Collaboration, E. Cortina Gil et al., Search for production of an invisible dark photon in π^0 decays. JHEP **05**, 182 (2019). arXiv:1903.08767 [hep-ex]. https://doi.org/10.1007/JHEP05(2019)182

52. BaBar Collaboration, J.P. Lees et al., Search for invisible decays of a dark photon produced in $e^+ e^-$ Collisions at BaBar. Phys. Rev. Lett. **119**(13), 131804 (2017). arXiv:1702.03327 [hep-ex]. https://doi.org/10.1103/PhysRevLett.119.131804

53. D. Banerjee et al., Dark matter search in missing energy events with NA64. Phys. Rev. Lett. **123**(12), 121801 (2019). arXiv:1906.00176 [hep-ex]. https://doi.org/10.1103/PhysRevLett.123.121801

54. Muon g-2 Collaboration, G.W. Bennett et al., Final report of the muon E821 anomalous magnetic moment measurement at BNL. Phys. Rev. **D73**, 072003 (2006). arXiv:hep-ex/0602035 [hep-ex]. https://doi.org/10.1103/PhysRevD.73.072003

55. KLEVER Project Collaboration, F. Ambrosino et al., KLEVER: an experiment to measure $BR(K_L \rightarrow \pi^0 \nu \bar{\nu})$ at the CERN SPS. arXiv:1901.03099 [hep-ex]

56. M. Raggi, V. Kozhuharov, P. Valente, The PADME experiment at LNF. EPJ Web Conf. **96**, 01025 (2015). https://doi.org/10.1051/epjconf/20159601025. arXiv:1501.01867 [hep-ex]

57. LDMX Collaboration, T. Akesson et al., Light Dark Matter eXperiment (LDMX). arXiv:1808.05219 [hep-ex]

58. T. Akesson et al., Dark Sector Physics with a Primary Electron Beam Facility at CERN. http://cds.cern.ch/record/2640784?ln=en

59. T. Raubenheimer, A. Beukers, A. Fry, C. Hast, T. Markiewicz, Y. Nosochkov, N. Phinney, P. Schuster, N. Toro, DASEL: Dark Sector Experiments at LCLS-II. arXiv:1801.07867 [physics.acc-ph]

60. A. Caputo, H. Liu, S. Mishra-Sharma, J.T. Ruderman, Dark Photon Oscillations in Our Inhomogeneous Universe. arXiv:2002.05165 [astro-ph.CO]

61. A.A. Garcia, K. Bondarenko, S. Ploeckinger, J. Pradler, A. Sokolenko, Effective photon mass and (dark) photon conversion in the inhomogeneous Universe. arXiv:2003.10465 [astro-ph.CO]

62. S.J. Witte, S. Rosauro-Alcaraz, S.D. McDermott, V. Poulin, Dark Photon Dark Matter in the Presence of Inhomogeneous Structure. arXiv:2003.13698 [astro-ph.CO]

63. S.D. McDermott, S. J. Witte, Cosmological evolution of light dark photon dark matter. Phys. Rev. D **101**(6), 063030 (2020). arXiv:1911.05086 [hep-ph]. https://doi.org/10.1103/PhysRevD.101.063030

64. K. Ehret et al., New ALPS results on hidden-sector lightweights. Phys. Lett. **B689** 149–155 (2010). arXiv:1004.1313 [hep-ex]. https://doi.org/10.1016/j.physletb.2010.04.066

65. M. Betz, F. Caspers, M. Gasior, M. Thumm, S. Rieger, First results of the CERN resonant weakly interacting sub-eV particle search (CROWS). Phys. Rev. D **88**(7), 075014 (2013). arXiv:1310.8098 [physics.ins-det]. https://doi.org/10.1103/PhysRevD.88.075014

66. J. Redondo, Helioscope bounds on hidden sector photons. JCAP **07**, 008 (2008). https://doi.org/10.1088/1475-7516/2008/07/008. arXiv:0801.1527 [hep-ph]

67. H. An, M. Pospelov, J. Pradler, Dark matter detectors as dark photon helioscopes. Phys. Rev. Lett. **111**, 041302 (2013). https://doi.org/10.1103/PhysRevLett.111.041302. arXiv:1304.3461 [hep-ph]

68. M. Schwarz, E.-A. Knabbe, A. Lindner, J. Redondo, A. Ringwald, M. Schneide, J. Susol, G. Wiedemann, Results from the Solar Hidden Photon Search (SHIPS). JCAP **08**, 011 (2015). https://doi.org/10.1088/1475-7516/2015/08/011. arXiv:1502.04490 [hep-ph]

69. M. Danilov, S. Demidov, D. Gorbunov, Constraints on hidden photons produced in nuclear reactors. Phys. Rev. Lett. **122**(4), 041801 (2019). arXiv:1804.10777 [hep-ph]. https://doi.org/10.1103/PhysRevLett.122.041801

70. J. Jaeckel, S. Roy, Spectroscopy as a test of Coulomb's law: a Probe of the hidden sector. Phys. Rev. D **82**, 125020 (2010). https://doi.org/10.1103/PhysRevD.82.125020. arXiv:1008.3536 [hep-ph]

71. J. Redondo, G. Raffelt, Solar constraints on hidden photons re-visited. JCAP **1308**, 034 (2013). https://doi.org/10.1088/1475-7516/2013/08/034. arXiv:1305.2920 [hep-ph]

72. H. An, M. Pospelov, J. Pradler, New stellar constraints on dark photons. Phys. Lett. **B725**, 190–195 (2013). arXiv:1302.3884 [hep-ph]. https://doi.org/10.1016/j.physletb.2013.07.008

73. H. An, M. Pospelov, J. Pradler, A. Ritz, Direct detection constraints on dark photon dark matter. Phys. Lett. B **747**, 331–338 (2015). arXiv:1412.8378 [hep-ph]. https://doi.org/10.1016/j.physletb.2015.06.018

74. XENON10 Collaboration, J. Angle et al., A search for light dark matter in XENON10 data. Phys. Rev. Lett. **107**, 051301 (2011). arXiv:1104.3088 [astro-ph.CO]. [Erratum: Phys.Rev.Lett. 110, 249901 (2013)]. https://doi.org/10.1103/PhysRevLett.107.051301

75. XENON100 Collaboration, E. Aprile et al., First Axion results from the XENON100 experiment. Phys. Rev. **D90**(6), 062009 (2014). arXiv:1404.1455 [astro-ph.CO]. [Erratum: Phys. Rev.D95,no.2,029904(2017)]. https://doi.org/10.1103/PhysRevD.90.062009, https://doi.org/10.1103/PhysRevD.95.029904

76. XENON Collaboration, E. Aprile et al., Light dark matter search with ionization signals in XENON1T. Phys. Rev. Lett. **123**(25), 251801 (2019). arXiv:1907.11485 [hep-ex]. https://doi.org/10.1103/PhysRevLett.123.251801

77. XENON Collaboration, E. Aprile et al., Observation of Excess Electronic Recoil Events in XENON1T. arXiv:2006.09721 [hep-ex]

78. DAMIC Collaboration, A. Aguilar-Arevalo et al., First direct-detection constraints on eV-Scale hidden-photon dark matter with DAMIC at SNOLAB. Phys. Rev. Lett. **118**, (14), 141803 (2017). arXiv:1611.03066 [astro-ph.CO]. https://doi.org/10.1103/PhysRevLett.118.141803

79. SuperCDMS Collaboration, T. Aralis et al., Constraints on dark photons and Axion-Like particles from SuperCDMS Soudan. Phys. Rev. **D101**(5), 052008 (2020). arXiv:1911.11905 [hep-ex]. https://doi.org/10.1103/PhysRevD.101.052008

80. Z. She et al., Direct detection constraints on dark photons with CDEX-10 experiment at the China Jinping underground laboratory. Phys. Rev. Lett. **124**(11), 111301 (2020). arXiv:1910.13234 [hep-ex]. https://doi.org/10.1103/PhysRevLett.124.111301

81. EDELWEISS Collaboration, E. Armengaud et al., Searches for electron interactions induced by new physics in the EDELWEISS-III Germanium bolometers. Phys. Rev. **D98**(8), 082004 (2018). arXiv:1808.02340 [hep-ex]. https://doi.org/10.1103/PhysRevD.98.082004

82. SENSEI Collaboration, O. Abramoff et al., SENSEI: direct-detection constraints on Sub-GeV dark matter from a shallow underground run using a prototype Skipper-CCD. Phys. Rev. Lett. **122**(16), 161801 (2019). arXiv:1901.10478 [hep-ex]. https://doi.org/10.1103/PhysRevLett. 122.161801

83. XMASS Collaboration, K. Abe et al., Search for dark matter in the form of hidden photons and axion-like particles in the XMASS detector. Phys. Lett. **B787**, 153–158 (2018). arXiv:1807.08516 [astro-ph.CO]. https://doi.org/10.1016/j.physletb.2018.10.050

84. FUNK Experiment Collaboration, A. Andrianavalomahefa et al., Limits from the Funk Experiment on the Mixing Strength of Hidden-Photon Dark Matter in the Visible and Near-Ultraviolet Wavelength Range. arXiv:2003.13144 [astro-ph.CO]

85. D.F. Bartlett, P.E. Goldhagen, E.A. Phillips, Experimental test of Coulomb's Law. Phys. Rev. D **2**, 483–487 (1970). https://doi.org/10.1103/PhysRevD.2.483

86. TEXONO Collaboration, M. Deniz et al., Measurement of Nu(e)-bar-Electron scattering cross-section with a CsI(Tl) scintillating crystal array at the Kuo-Sheng nuclear power reactor. Phys. Rev. D **81**, 072001 (2010). arXiv:0911.1597 [hep-ex]. https://doi.org/10.1103/PhysRevD.81.072001

87. D. Fixsen, E. Cheng, J. Gales, J.C. Mather, R. Shafer, E. Wright, The cosmic microwave background spectrum from the full COBE FIRAS data set. Astrophys. J. **473**, 576 (1996). https://doi.org/10.1086/178173. arXiv:astro-ph/9605054

88. A. Mirizzi, J. Redondo, G. Sigl, Microwave background constraints on mixing of photons with hidden photons. JCAP **03**, 026 (2009). https://doi.org/10.1088/1475-7516/2009/03/026. arXiv:0901.0014 [hep-ph]

89. D. Wadekar, G.R. Farrar, First direct astrophysical constraints on dark matter interactions with ordinary matter at very low velocities. arXiv:1903.12190 [hep-ph]

90. A. Bhoonah, J. Bramante, F. Elahi, S. Schon, Galactic Center gas clouds and novel bounds on ultralight dark photon, vector portal, strongly interacting, composite, and super-heavy dark matter. Phys. Rev. D **100**(2), 023001 (2019). arXiv:1812.10919 [hep-ph]. https://doi.org/10.1103/PhysRevD.100.023001

91. P. Arias, D. Cadamuro, M. Goodsell, J. Jaeckel, J. Redondo, A. Ringwald, WISPy cold dark matter. JCAP **1206**, 013 (2012). https://doi.org/10.1088/1475-7516/2012/06/013. arXiv:1201.5902 [hep-ph]

92. J. Redondo, M. Postma, Massive hidden photons as lukewarm dark matter. JCAP **02**, 005 (2009). https://doi.org/10.1088/1475-7516/2009/02/005. arXiv:0811.0326 [hep-ph]

93. P. Sikivie, Experimental tests of the invisible Axion. Phys. Rev. Lett. **51**, 1415–1417 (1983). [Erratum: Phys.Rev.Lett. 52, 695 (1984)]. https://doi.org/10.1103/PhysRevLett.51.1415

94. P. deNiverville, M. Pospelov, A. Ritz, Observing a light dark matter beam with neutrino experiments. Phys. Rev. D **84**, 075020 (2011). arXiv:1107.4580 [hep-ph]. https://doi.org/10.1103/PhysRevD.84.075020

95. MiniBooNE DM Collaboration, A. Aguilar-Arevalo et al., Dark matter search in nucleon, pion, and electron channels from a proton beam dump with MiniBooNE. Phys. Rev. D **98**(11), 112004 (2018). arXiv:1807.06137 [hep-ex]. https://doi.org/10.1103/PhysRevD.98.112004

96. CRESST Collaboration, G. Angloher et al., Results on light dark matter particles with a low-threshold CRESST-II detector. Eur. Phys. J. C **76**(1), 25 (2016). arXiv:1509.01515 [astro-ph.CO]. https://doi.org/10.1140/epjc/s10052-016-3877-3

97. SHiP Collaboration, M. Anelli et al., A facility to Search for Hidden Particles (SHiP) at the CERN SPS. arXiv:1504.04956 [physics.ins-det]

98. BDX Collaboration, M. Battaglieri et al., Dark Matter Search in a Beam-Dump eXperiment (BDX) at Jefferson Lab. arXiv:1607.01390 [hep-ex]

99. MicroBooNE, LAr1-ND, ICARUS-WA104 Collaboration, M. Antonello et al., A Proposal for a Three Detector Short-Baseline Neutrino Oscillation Program in the Fermilab Booster Neutrino Beam. arXiv:1503.01520 [physics.ins-det]

100. M. Battaglieri et al., US Cosmic Visions: New Ideas in Dark Matter 2017: Community Report in U.S. Cosmic Visions: New Ideas in Dark Matter. 7, 2017. https://doi.org/10.1016/j.physletb. 2020.135258. arXiv:1707.04591 [hep-ph]

101. SuperCDMS Collaboration, R. Agnese et al., Projected sensitivity of the SuperCDMS SNO-LAB experiment. Phys. Rev. D **95**(8), 082002 (2017). arXiv:1610.00006 [physics.ins-det]. https://doi.org/10.1103/PhysRevD.95.082002

102. Planck Collaboration, P. Ade et al., Planck 2015 results. XIII. Cosmological parameters. Astron. Astrophys. **594**, A13 (2016). arXiv:1502.01589 [astro-ph.CO]. https://doi.org/10.1051/0004-6361/201525830

103. A. Caputo, A. Esposito, E. Geoffray, A.D. Polosa, S. Sun, Dark matter, dark photon and superfluid He-4 from effective field theory. Phys. Lett. B **802**, 135258 (2020). https://doi.org/10.1016/j.physletb.2020.135258. arXiv:1911.04511 [hep-ph]

104. SENSEI Collaboration, J. Tienberg, M. Sofo-Haro, A. Drlica-Wagner, R. Essig, Y. Guardincerri, S. Holland, T. Volansky, and T.-T. Yu, Single-electron and single-photon sensitivity with a silicon Skipper CCD. Phys. Rev. Lett. **119**(13), 131802 (2017). arXiv:1706.00028 [physics.ins-det]

105. L. Barak et al., SENSEI: Direct-Detection Results on sub-GeV Dark Matter from a New Skipper-CCD. arXiv:2004.11378 [astro-ph.CO]

Chapter 4
Outlook

In the past 50 years it has been assumed that physics beyond the SM interacted through (at least) some of the same gauge interactions of the SM. The minimal supersymmetric SM and weakly interacting massive dark matter are the two preeminent and most influential models based on this paradigm.

This program is now running out of some of the initial momentum because of the lack of the discovery of new particles. In the absence of new states, the many parameters, for instance, of the minimal supersymmetric SM are working against its usefulness as a foil for the SM in mapping possible experimental discrepancies.

In more recent times—mostly under the influence of this lack of any real signal of the breaking up of the SM—a more general scenario has been attracting increasing interest. Matter beyond the SM is part of a new sector which is dark because it does not interact through the SM gauge interactions. The dark sector may contain a wealth of physics with many particles (some of which are dark matter) and interactions.

From our side, in the visible world, we may glimpse this dark sector through a portal. If it exists, this portal can take various forms depending on the spin of the mediator. We have reviewed in this primer the vector case in which the portal arises from the kinetic mixing between the SM electric (or hyper) charge gauge group and an $U(1)$ gauge symmetry of the dark sector.

The discovery of the dark photon associated to this new Abelian gauge symmetry is by far more interesting than finding just a new particle because, if found, this new gauge boson would be the harbinger of a new interaction and of the existence of a whole new sector of elementary particles.

Past and current experiments have already restricted an important part of the space of the parameters of the vector portal, both for the massless and the massive dark photon. Compared to other searches for models beyond the SM, the parameters are fewer and the signatures more easily interpreted.

We are now on the verge of a new wave of experiments aiming at further closing the windows left still open in the interaction between ordinary matter and the dark photon. The long years of searching in vane for physics beyond the SM has thought us to be very patient and persevering.

© The Author(s), under exclusive license to Springer Nature Switzerland AG 2021
M. Fabbrichesi et al., *The Physics of the Dark Photon*,
SpringerBriefs in Physics, https://doi.org/10.1007/978-3-030-62519-1_4

The constraints in the massless case seem to relegate the possible detection of the dark photon to very large values of the effective scale Λ in the dark dipole interaction, as we discuss in Sect. 2.1. Exploring physics at such a large energy scale requires the high sensitivity that can only be achieved either in future lepton colliders (where the scaling with the energy of the dark dipole operator will also enhance its contribution) or in searches for rare flavor-changing decays like those of the Kaon and B-meson systems.

The constraints in the case of the massive dark photon have left open two important regions in the parameter space. The first one is for the visible dark photon with masses around 100 MeV or larger and mixing parameter between 10^{-6} and 10^{-4}. Many future experiments aim at looking into this range, as we review in Sect. 3.3.1. If also this window will be closed, it means that the already feeble interaction of the vector portal is very weak indeed. Which leaves us with the second window still left unexplored: an invisible dark photon with a very light mass and a mixing parameter of order $O(10^{-8})$ or even lighter and with smaller mixing parameter, as discussed in Sects. 3.3.2 and 3.3.3. These two latter regions are of great interest for astrophysics and cosmology and a very active area of speculations.

No single experiment or experimental approach is sufficient alone to cover the large parameter space in terms of masses and couplings that dark photon models suggest: Synergy and complementarity among a great variety of experimental facilities are paramount, calling for a broad collaboration across different communities.

Chapter 5
Appendix

5.1 Dark Sector Portals

In this appendix we give the Lagrangian for the four portals mentioned in the introduction as well as a minimal bibliography to provide their context within the search
of phenomenological models beyond the Standard Model.

The dark sector is assumed to interact with the visible, SM sector through relevant operators of dimension four and five (and possibly sub-leading higher-order
operators).

These portals are classified according to the spin of the mediator field. We can
have

- dark photon (spin 1): The portal operator arises from the kinetic is mixing between
  the SM photon field strength $F_{\mu\nu}$ and a dark photon $F^{\mu\nu\prime}$:

$$\frac{\varepsilon}{2} F_{\mu\nu} F^{\mu\nu\prime} ,$$

it is an operator of dimension four. It is assumed that the dark photon is the main
carrier of the interaction among the dark sector states.
The existence of an independent $U(1)$ group symmetry was originally proposed,
in the context of supersymmetric theories, in [1, 2] and, more in general, in [3, 4];
- axion (spin 0): The operator comes from the interaction between a pseudo-scalar,
  the axion a, and the SM photon and fermions ψ:

$$\frac{a}{f_a} F_{\mu\nu} \tilde{F}^{\mu\nu} + \frac{1}{f_a} \partial_\mu a \, \overline{\psi} \gamma^\mu \gamma_5 \psi ,$$

with operators of dimension five; the physics of this portal is based on that of
the axion and related to the strong CP problem as well as axion dark matter. In
many cases, the portal is generalized to an axion-like particle (ALP) with similar

© The Author(s), under exclusive license to Springer Nature Switzerland AG 2021
M. Fabbrichesi et al., *The Physics of the Dark Photon*,
SpringerBriefs in Physics, https://doi.org/10.1007/978-3-030-62519-1_5

couplings but without the constraints of the QCD axion. The parameters of the portal are two: the mass m_a of the axion, or the ALP and the scale f_a. Often the ALP is the only member of the dark sector of these models.

The original axion emerged from addressing [5] the strong CP problem induced by instantons in QCD. Light pseudo-scalar bosons are found in many models of physics beyond the Standard Model;

• scalar (spin 0): Interaction between a scalar S and the SM Higgs boson H:

$$(\mu S + \lambda S^2) H^\dagger H ,$$

in this case the operators are of dimension three and four. The experimental limits are often expressed in terms of the two parameters ν and the mass m_S of the scalar singlet, and neglecting the quartic coupling λ. In most models, the dark sector states have Yukawa-like interactions with the scalar S.

The idea of a scalar singlet interacting with the Higgs boson originated within the framework of the next-to-minimal supersymmetric Standard Model [6–8] and developed independently in [9–11];

• sterile neutrino (spin 1/2): Interaction between a heavy fermion N, which is a SM singlet, the SM Higgs boson and the SM fermions L:

$$y_N \overline{L} H N ,$$

with, again, an operator of dimension four. The existence of heavy lepton-like fermions is suggested by neutrino see-saw models and the possible origin of baryon-number asymmetry in the leptonic sector. The experimental searches are framed in terms of the parameter y_N and the mass of the heavy fermion N. The sterile neutrino can be the only member of the dark sector or be one among many other dark fermions.

The structure of the neutrino portal closely follows that of the see-saw mechanism [12–15]—which was introduced to generate small masses for the neutrinos— and, more in general, left-right symmetric models [16–19].

More details on the various portals can be found in the same references cited in the introduction: [20–27].

5.2   Boltzmann Equation and Relic Density

This appendix includes a short summary of some results necessary to follow the discussion in the main text about the relic density of dark matter and limits based on cosmology. We follow the excellent review [28].

The rate Γ for a the interaction between two particles is given as

$$\Gamma = n \, \sigma \, v , \tag{5.1}$$

the product of the corresponding cross section σ times the number density of the particles partaking n, times the their relative velocity v.

This process proceeds as long as the rate is larger than the Hubble constant

$$H(T) = \frac{\pi\sqrt{g_*(T)}}{\sqrt{90}} \frac{T^2}{m_{Pl}} , \qquad (5.2)$$

where m_{Pl} the Planck mass and $g_*(T)$ is the number of effective degrees of freedom at the given temperature is given by

$$g_*(T) = \sum_{\text{bosons}} g_b \left(\frac{T_b}{T}\right)^4 + \frac{7}{8} \sum_{\text{fermions}} g_f \left(\frac{T_f}{T}\right)^4 , \qquad (5.3)$$

where $g_{b,f}$ is the number of degrees of freedom of the corresponding particle. The value of the function $g_*(T)$ goes from 106.5 above the EW phase transition to 3.38 at temperature around 0.1 MeV.

After $\Gamma < H$, the particles are decoupled and their number density frozen.

The number density at the equilibrium at a given temperature T (for $k_B = 1$) is given by

$$n_{eq}(T) = g_* \int \frac{d^3p}{(2\pi)^3} \frac{1}{e^{E/T} \pm 1}$$

$$= \begin{cases} g_* \left(\frac{mT}{2\pi}\right)^{3/2} e^{-m/T} & \text{non-relativistic } (T \ll m) \\[2mm] \frac{\zeta(3)}{\pi^2} g_* T^3 & \text{relativistic bosons } (T \gg m) \\[2mm] \frac{3}{4} \frac{\zeta(3)}{\pi^2} g_* T^3 & \text{relativistic fermions } (T \gg m) , \end{cases} \qquad (5.4)$$

where $\zeta(3) \simeq 1.2$

The number density $n(t)$ of a weakly interacting, massive particle χ at a certain time t in the evolution of the Universe is computed by means of the Boltzmann equation

$$\dot{n}(t) + 3 H(t) n(t) = -\langle \sigma_{\chi\chi \to ff} v\rangle \left(n^2(t) - n_{eq}^2(t)\right) , \qquad (5.5)$$

where $H(t)$ is the Hubble constant and $\langle \sigma_{\chi\chi \to ff} v\rangle$ is the thermal average of the cross section for a pair of the particles χ, with relative velocity $v = (s - 4m_\chi^2)/m_\chi^2$, to annihilate into SM fermions f; this term depletes the density as the particles χ turns into SM fermions. The thermal average is defined as

$$\langle \sigma_{\chi\chi \to ff} v\rangle = \frac{\int_\infty^{4m_\chi^2} ds \sqrt{s}(s - 4m_\chi^2) K_1\left(\frac{\sqrt{s}}{T}\right) \sigma_{\chi\chi \to ff}}{8 m_\chi^4 T \left[K_2\left(\frac{m_\chi}{T}\right)\right]^2} , \qquad (5.6)$$

where K_1 and K_2 are the Bessel function of second kind. It is usually computed after expanding

$$\langle \sigma_{\chi\chi \to ff} v \rangle = \langle s_0 + s_1 v^2 + O(v^4) \rangle \tag{5.7}$$

with s_0 the cross section in the s-wave and s_1 the first correction in the p-wave. The leading term is s_0 for the dark sector Dirac fermions interacting through the dark photon, in both the s- and t-channel.

Equation (5.5) is usually re-written in terms of the function $Y(t) = n(t)/T^3$ and the variable $x = m_\chi/T = \sqrt{2t H(T = m_\chi)}$ as

$$\frac{dY}{dx} = -\frac{\lambda(x)}{x^2} \left[Y^2(x) - Y_{eq}^2 \right] \tag{5.8}$$

with

$$\lambda(x) = \frac{m_\chi^3 \langle \sigma_{\chi\chi \to ff} v \rangle}{H(T = m_\chi)}, \tag{5.9}$$

and in this form numerically solved.

Equation (5.8) can be solved analytically by dropping the second term $Y_{eq}^2(x)$— which is small because decreasing like e^{-x}—approximating

$$\langle \sigma_{\chi\chi \to ff} v \rangle = \sigma_{\chi\chi \to ff} v + O(v^2), \tag{5.10}$$

where $v = \sqrt{2/x}$ and writing

$$\lambda(x) = \frac{\sqrt{180}\, m_{Pl}\, m_\chi}{\pi \sqrt{g_* x}} \sigma_{\chi\chi \to ff} \tag{5.11}$$

by means of

$$H(T = m_\chi) = \frac{\pi \sqrt{g_*(T = m_\chi)}}{90} \frac{m_\chi^2}{m_{Pl}}. \tag{5.12}$$

The solution for x' larger than decoupling temperature x_d is

$$Y(x') = \frac{x_d}{\lambda}. \tag{5.13}$$

This quantity is related to the relic density

$$\rho_\chi = m_\chi n(x') = m_\chi^4 \frac{Y(x')}{28 x_d} \tag{5.14}$$

or, in terms of the normalized quantity $\Omega_\chi = \rho_\chi/\rho_c$ as

$$\Omega_\chi h^2 \simeq 0.12 \frac{x_d}{23} \frac{\sqrt{g_*}}{10} \frac{1.7 \times 10^{-9} \text{GeV}^{-2}}{\langle \sigma_{\chi\chi \to ff} v \rangle}$$

$$\simeq \frac{2.5 \times 10^{-10} \text{ GeV}^{-2}}{\langle \sigma_{\chi\chi \to ff} v \rangle}, \tag{5.15}$$

which provides the relationship between relic density and the annihilation cross section.

5.3 Thermal Field Theory

The energy loss rate $\mathcal{Q}$ (energy per volume and unit time) for the emission of a pseudoscalar particle (the axion) in a process with matrix element, which is computed in the vacuum, is given by

$$\mathcal{Q} = \prod_{i=1} \int \frac{d^3 \mathbf{p}_i}{2E_i (2\pi)^3} f_i(E_i) \prod_{f=1} \int \frac{d^3 \mathbf{p}_f}{2E_f (2\pi)^3} \left[1 \pm f_f(E_f)\right] \int \frac{d^3 \mathbf{p}_a}{2\omega_a (2\pi)^3} \omega_a$$

$$\times \frac{1}{\mathcal{S}} \sum_{\text{spin and pol.}} |\mathcal{M}|^2 (2\pi)^4 \delta^4 \left(\sum p_i - \sum p_f - p_a\right), \tag{5.16}$$

where $\mathcal{S}$ is a symmetrization factor for identical particles. In Eq. (5.16), the medium is composed of the initial particles i and final particles f with the corresponding energy ω and momentum $\mathbf{p}$ and with occupation number following the distribution function (Fermi or Bose depending on the particles) ($k_B = 1$):

$$n_j(E_j) = g_j \int \frac{d^2 \mathbf{p}_j}{(2\pi)^3} f(E_j), \tag{5.17}$$

where g_j is the degeneracy number. The emitted axion carries energy ω_a and momentum $\mathbf{p}_a$.

Given the squared matrix element $\sum |\mathcal{M}|^2$ for the process of interest, the electron and nucleon *Bremsstrahlung* in Sect. 2.1, the corresponding luminosity can be computed as

$$L = \int dV \mathcal{Q} e^{-\tau}, \tag{5.18}$$

where τ is an attenuation factor taking into account the optical depth of the emission, and compared to the observational data.

When the emitted particle mixes with the ordinary photon, the approach above of computing the matrix element in the vacuum is no longer a reliable approximation and the full thermal field theory must be used. We follow [29] in giving the essential formulas.

The electromagnetic polarization tensor is given by

$$\Pi^{\mu\nu}(k) = 16\pi\alpha \int \frac{d^3\mathbf{p}_i}{(2\pi)^3} \frac{1}{2E} [n_e(E) + n_{\bar{e}}(E)]$$
$$\times \frac{p\cdot k\,(p^\mu k^\nu + k^\mu p^\nu) - k^2 p^\mu p^\nu - (p\cdot k)^2 g^{\mu\nu}}{(p\cdot k)^2 - (k^2)^2/4}, \tag{5.19}$$

where $k^\mu = (\omega, \mathbf{k})$ and $p^\mu = (E, \mathbf{p})$. The transverse and longitudinal polarizations are defined as

$$\Pi_T(\omega, \mathbf{k}) = \frac{1}{2}\left(\delta^{ij} - k^i k^j\right) \Pi^{ij}(\omega, \mathbf{k}) \tag{5.20}$$

and

$$\Pi_L(\omega, \mathbf{k}) = \Pi^{00}(\omega, \mathbf{k}). \tag{5.21}$$

The effective propagator of the photon (in the Coulomb gauge) has components

$$D^{00}(\omega, k) = \frac{1}{k^2 - \Pi_L(\omega, k)} \tag{5.22}$$

and

$$D^{ij}(\omega, k) = \frac{1}{k^2 - \Pi_T(\omega, k)}\left(\delta^{ij} - k^i k^j\right). \tag{5.23}$$

The dispersion relationships are defined by the solutions of the equations

$$\omega_T^2 = k^2 + \Pi_T(\omega_T, k) \quad \text{and} \quad \omega_L^2 = \frac{\omega_L^2}{k^2} + \Pi_T(\omega_L, k) \tag{5.24}$$

In the degenerate limit, the distribution functions in Eq. (5.17) reduce to step functions at the Fermi momentum $p_F = \sqrt{3\pi^2 n_e}$ and we have

$$\Pi_T(\omega, \mathbf{k}) = \omega_P^2 \frac{3\omega^2}{2v_F^2 k^2}\left(1 - \frac{\omega^2 - v_F^2 k^2}{2v_F\omega k}\log\frac{\omega + v_F k}{\omega - v_F k}\right) \tag{5.25}$$

and

$$\Pi_L(\omega, \mathbf{k}) = \omega_P^2 \frac{3\omega}{2v_F^3 k}\left(\frac{\omega}{2v_F k}\log\frac{\omega + v_F k}{\omega - v_F k} - 1\right) \tag{5.26}$$

where $\omega_P = 4\alpha p_F^2 v_F/3\pi$ is the plasma frequency.

The energy loss rate Q (energy per volume and unit time) is written in terms of the imaginary part of the polarization of the photon in the medium of charged particles. The contribution of the longitudinal and transverse modes is obtained (by the optical theorem) as

$$\mathcal{Q} = -\int \frac{d^3\mathbf{k}}{(2\pi)^3} \frac{\operatorname{Im}\Pi_L(\omega, \mathbf{k}) + \operatorname{Im}\Pi_T(\omega, \mathbf{k})}{\omega(e^{\omega/T} - 1)}, \tag{5.27}$$

from which the luminosity in Eq. (5.18) can be computed and compared with the astrophysical limit of interest. The expression for $\operatorname{Im}\Pi_{L,T}(\omega, \mathbf{k})$ for the massive dark photon can be found in [30–32] where plasma effects are also included.

References

1. P. Fayet, Effects of the Spin 1 partner of the Goldstino (Gravitino) on neutral current phenomenology. Phys. Lett. B **95**, 285–289 (1980)
2. P. Fayet, On the search for a new Spin 1 Boson. Nucl. Phys. B **187**, 184–204 (1981)
3. L.B. Okun, Limits of electrodynamics: paraphotons? Sov. Phys. JETP **56**, 502 (1982). [Zh. Eksp. Teor. Fiz.83,892(1982)]
4. H. Georgi, P.H. Ginsparg, S.L. Glashow, Photon oscillations and the cosmic background radiation. Nature **306**, 765–766 (1983)
5. R. Peccei, H.R. Quinn, CP conservation in the presence of instantons. Phys. Rev. Lett. **38**, 1440–1443 (1977)
6. J.R. Ellis, J. Gunion, H.E. Haber, L. Roszkowski, F. Zwirner, Higgs bosons in a nonminimal supersymmetric model. Phys. Rev. D **39**, 844 (1989)
7. J. Derendinger, C.A. Savoy, Quantum effects and SU(2) x U(1) breaking in supergravity gauge theories. Nucl. Phys. B **237**, 307–328 (1984)
8. J. Frere, D. Jones, S. Raby, Fermion masses and induction of the weak scale by supergravity. Nucl. Phys. B **222**, 11–19 (1983)
9. V. Silveira, A. Zee, Scalar phantoms. Phys. Lett. B **161**, 136–140 (1985)
10. T. Binoth, J. van der Bij, Influence of strongly coupled, hidden scalars on Higgs signals. Z. Phys. C **75**, 17–25 (1997). arXiv:hep-ph/9608245
11. B. Patt, F. Wilczek, Higgs-field portal into hidden sectors. arXiv:hep-ph/0605188
12. P. Minkowski, $\mu \to e\gamma$ at a Rate of One Out of 10^9 Muon Decays? Phys. Lett. B **67**, 421–428 (1977)
13. T. Yanagida, Horizontal symmetry and masses of neutrinos. Prog. Theor. Phys. **64**, 1103 (1980)
14. M. Gell-Mann, P. Ramond, R. Slansky, Complex spinors and unified theories. Conf. Proc. C **790927**, 315–321 (1979). arXiv:1306.4669 [hep-th]
15. R.N. Mohapatra, G. Senjanovic, Neutrino masses and mixings in gauge models with spontaneous parity violation. Phys. Rev. D **23**, 165 (1981)
16. J.C. Pati, A. Salam, Lepton number as the fourth color. Phys. Rev. D **10**, 275–289 (1974). [Erratum: Phys.Rev.D 11, 703–703 (1975)]
17. R. Mohapatra, J.C. Pati, A natural left-right symmetry. Phys. Rev. D **11**, 2558 (1975)
18. R.N. Mohapatra, J.C. Pati, Left-Right gauge symmetry and an Isoconjugate model of CP violation. Phys. Rev. D **11**, 566–571 (1975)
19. G. Senjanovic, R.N. Mohapatra, Exact left-right symmetry and spontaneous violation of parity. Phys. Rev. D **12**, 1502 (1975)
20. J.L. Hewett et al., Fundamental Physics at the Intensity Frontier (2012). arXiv:1205.2671 [hep-ex]. http://lss.fnal.gov/archive/preprint/fermilab-conf-12-879-ppd.shtml
21. R. Essig et al., Working group report: new light weakly coupled particles, in *Proceedings, 2013 Community Summer Study on the Future of U.S. Particle Physics: Snowmass on the Mississippi (CSS2013): Minneapolis, MN, USA*, July 29–Aug 6, 2013. arXiv:1311.0029 [hep-ph]. http://www.slac.stanford.edu/econf/C1307292/docs/IntensityFrontier/NewLight-17.pdf
22. M. Raggi, V. Kozhuharov, Results and perspectives in dark photon physics. Riv. Nuovo Cim. **38**(10), 449–505 (2015)

23. M.A. Deliyergiyev, Recent progress in search for dark sector signatures. Open Phys. **14**(1), 281–303 (2016). arXiv:1510.06927 [hep-ph]
24. S. Alekhin et al., A facility to search for hidden particles at the CERN SPS: the SHiP physics case. Rept. Prog. Phys. **79**(12), 124201 (2016). arXiv:1504.04855 [hep-ph]
25. F. Curciarello, Review on dark photon. EPJ Web Conf. **118**, 01008 (2016)
26. J. Alexander et al., Dark Sectors 2016 Workshop: Community Report (2016). arXiv:1608.08632 [hep-ph]. http://lss.fnal.gov/archive/2016/conf/fermilab-conf-16-421.pdf
27. J. Beacham et al., Physics beyond colliders at CERN: beyond the standard model working group report. J. Phys. G **47**(1), 010501 (2020). arXiv:1901.09966 [hep-ex]
28. M. Bauer, T. Plehn, Yet another introduction to dark matter. Lect. Notes Phys. **959** (2019). arXiv:1705.01987 [hep-ph]
29. E. Braaten, D. Segel, Neutrino energy loss from the plasma process at all temperatures and densities. Phys. Rev. D **48**, 1478–1491 (1993). arXiv:hep-ph/9302213
30. H. An, M. Pospelov, J. Pradler, New stellar constraints on dark photons. Phys. Lett. **B725**, 190–195 (2013). arXiv:1302.3884 [hep-ph]
31. J. Redondo, G. Raffelt, Solar constraints on hidden photons re-visited. JCAP **1308**, 034 (2013). arXiv:1305.2920 [hep-ph]
32. E. Hardy, R. Lasenby, Stellar cooling bounds on new light particles: plasma mixing effects, ccc. JHEP **02**, 033 (2017). arXiv:1611.05852 [hep-ph]

Printed in the United States
By Bookmasters